Solar Radiation - Measurement, Modeling and Forecasting Techniques for Photovoltaic Solar Energy Applications

Edited by Mohammadreza Aghaei

Published in London, United Kingdom

IntechOpen

Supporting open minds since 2005

Solar Radiation - Measurement, Modeling and Forecasting Techniques for Photovoltaic Solar Energy Applications
http://dx.doi.org/10.5772/intechopen.87671
Edited by Mohammadreza Aghaei

Contributors
Mohammadreza Aghaei, Marc Korevaar, Pavel Babal, Hesan Ziar, Olfa Bel Hadj Brahim Kechiche, Habib Sammouda, Ismail Kaaya, Julian Ascencio-Vasquez, Abderrahmen Ben Chaabene, Khira Ouelhazi, Ines Sansa, Najiba Mrabet Bellaaj, Chetna Ameta, Reema Agarwal, Yogeshwari Vyas, Priyanka Chundawat, Dharmendra, David Afungchui, Joseph Ebobenow, Ali Helali, Nkongho Ayuketang Arreyndip, Rudy Calif, Maina André, Mohammad Aminul Islam, Nabilah M. Kassim, Marc A.N. Korevaar, Amir Nedaei, Aref Eskandari, Jafar Milimonfared

Notice
Statements and opinions expressed in the chapters are these of the individual contributors and not necessarily those of the editors or publisher. No responsibility is accepted for the accuracy of information contained in the published chapters. The publisher assumes no responsibility for any damage or injury to persons or property arising out of the use of any materials, instructions, methods or ideas contained in the book.

First published in London, United Kingdom, 2022 by IntechOpen
IntechOpen is the global imprint of INTECHOPEN LIMITED, registered in England and Wales, registration number: 11086078, 5 Princes Gate Court, London, SW7 2QJ, United Kingdom
Printed in Croatia

British Library Cataloguing-in-Publication Data
A catalogue record for this book is available from the British Library

Additional hard and PDF copies can be obtained from orders@intechopen.com

Solar Radiation - Measurement, Modeling and Forecasting Techniques for Photovoltaic Solar Energy Applications
Edited by Mohammadreza Aghaei
p. cm.
Print ISBN 978-1-83968-858-4
Online ISBN 978-1-83968-859-1
eBook (PDF) ISBN 978-1-83968-860-7

We are IntechOpen,
the world's leading publisher of
Open Access books
Built by scientists, for scientists

6,000+
Open access books available

148,000+
International authors and editors

185M+
Downloads

Our authors are among the

156
Countries delivered to

Top 1%
most cited scientists

12.2%
Contributors from top 500 universities

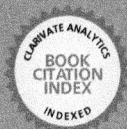

Interested in publishing with us?
Contact book.department@intechopen.com

Numbers displayed above are based on latest data collected.
For more information visit www.intechopen.com

Meet the editor

Mohammadreza Aghaei is a senior researcher and lecturer in the field of renewables, photovoltaics, enabling technologies, and energy systems. He received a Ph.D. in Electrical Engineering from Politecnico di Milano, Italy in 2016. He was a postdoctoral scientist at Fraunhofer ISE and Helmholtz-Zentrum Berlin-PV-comB, Germany, in 2017 and 2018, respectively. He has been a guest scientist at the University of Freiburg, Germany since 2017. He completed two years of postdoc studies at Eindhoven University of Technology, Netherlands. Currently, he is a senior scientist at the Norwegian University of Science and Technology and co-leader of an EU H2020-funded project "COLLECTiEF." Dr. Aghaei has been appointed as an adjunct professor at the Amirkabir University of Technology, Iran. He is also a senior lecturer at the University of Freiburg. He was the chair of COST Action PEARL PV's Working Group 2: Reliability and Durability of PV. Dr. Aghaei is a senior member of the Institute of Electrical and Electronics Engineers (IEEE). He has a very strong international profile and more than 10 years of research experience in renewable energy systems. He was involved in several national and EU-funded research projects dealing with photovoltaics, energy systems, and smart and autonomous technology. He has authored numerous publications in internationally refereed journals, book chapters, and conference proceedings. He has also supervised several master's and Ph.D. students.

Contents

Preface

Energy consumption across the world is increasing, and according to International Energy Agency (IEA) reports, humanity will require much more energy in the upcoming decades. Therefore, the world will be faced with an imminent energy dilemma in the 21st century.

The energy transition is a pathway toward the transformation of the global energy sector from fossil-based fuels to zero-carbon emissions. All energy scenarios project that global energy generation and consumption will increase by more than 50% by 2050. This means the global energy sectors face greater uncertainty and challenges in both the short and long term. The decarbonization of the energy sector requires urgent action on a global scale to reduce carbon emissions and mitigate the effects of climate change.

To overcome these challenges, humankind needs to harness the power of the sun as an infinite source to supply additional energy as well as increase the share of renewable energy around the world. Solar radiation is radiant energy originating from the sun in the form of electromagnetic radiation at various wavelengths. Almost all renewable energy comes from the sun either directly or indirectly. A vast amount of solar energy (173,000 terawatts) reaches the atmosphere and surface of the Earth, which is more than 10,000 times greater than the total energy used in the world. Today, photovoltaic (PV) solar energy has become the cheapest source for electrical power generation. At the beginning of 2022, PV installation exceeded 1 TW, which was an impressive milestone in the solar energy sector. In 2021, the world installed at least 183 GW, and PV capacity reached 788 GW at the end of 2020.

This book provides detailed information about solar radiation as the source of PV solar energy. It addresses various technical and practical aspects, including fundamental principles, measurement, modeling, and forecasting of solar radiation for PV solar energy technologies and applications. Most of this book describes the basic, modern, and contemporary knowledge and technology of extraterrestrial and terrestrial solar irradiance for PV solar energy. The contents contribute to energy transition and the United Nations' Sustainable Development Goals (SDGs) directly (SDG7: Affordable and Clean Energy; SDG13: Climate Action) and indirectly (SDG8: Decent Work and Economic Growth; SDG9: Industry, Innovation, and Infrastructure; SDG11: Sustainable Cities and Communities).

The book includes eleven chapters categorized into four sections: (I) "Introduction," (II) "Fundamentals, Measurements and Modeling of Solar Radiation", (III) "Forecasting and Characterization of Solar Radiation," and (IV) "Solar Photovoltaic Technologies and Applications."

Section I includes Chapter 1, which introduces the concept of energy transition and presents the background of solar radiation and solar energy as well as provides an overview of technologies, applications, and trends of solar photovoltaics. As an introductory chapter, it also reports statistical information on the status of photovoltaics from global installation, recent developments, and pioneer countries to the largest installed PV plants and future perspectives.

Section II includes Chapters 2 and 3, which present detailed information on the measurement and modeling of solar irradiance for solar PV energy. Chapter 2, "Measuring Solar Irradiance for Photovoltaics", discusses the characteristics and different components of solar irradiance and the instruments for measurement of these components. It gives detailed information on the physics involved in the measurement instruments and their calibration and corresponding uncertainty. Chapter 3, "Modelling of Solar Radiation for Photovoltaic Applications", quantifies different models in which solar radiation can be used in PV applications. It also presents various linear and non-linear solar radiation models that incorporate different combinations of parameters, namely, clearness index, the sunshine fraction, cloud cover, and air mass. The given models aim to estimate the direct and diffuse components of global solar radiation on both the horizontal and tilted surfaces to determine the optimal tilt and azimuthal angles for solar PV applications.

Section III consists of Chapters 4 and 5, which summarize forecasting and characterization methods for solar radiation to improve the performance of PV systems. Chapter 4, "Forecasting and Modelling of Solar Radiation for Photovoltaic (PV) Systems", presents a time series method for the prediction of solar radiation using the Auto-Regressive and Moving model, resulting in PV power forecasting. Chapter 5, "Temporal Fluctuations Scaling Analysis: Power Law of Ramp Rate's Variance for PV Power Output", focuses on the quantification of ramp rate's variance at different short time scales for tropical measurement sites that exhibit high irradiance variability due to complex microclimatic context. The outcome of this study is based on a statistical perspective in the solar PV energy area that introduces the multifractality analysis of variability of PV power output during the daytime.

Section IV contains Chapters 6–11, which cover solar PV technologies and applications ranging from solar cells, reliability assessment, outdoor characterization, and conventional and emerging PV technologies to bifacial PV technology, PV power prediction, concentrator PV system, and novel control methods for maximizing the PV system output. Chapter 6, "Assessing the Impact of Spectral Irradiance on the Performance of Different Photovoltaic Technologies", discusses different commercially available technologies of PV cells including crystalline silicon (c-Si), polycrystalline silicon (pc-Si), cadmium telluride (CdTe), and copper indium gallium selenide (CIGS). It presents a correlation study on the spectral response or the photocurrent of different PV cells with the variations of the solar spectrum, environmental conditions, and the material properties and construction of PV cells. Chapter 7, "Outdoor Performance and Stability Assessment of Dye-Sensitized Solar Cells (DSSCs)", discusses the principle of dye-sensitized solar cells and studies the outdoor performance and long-term stability of dye-sensitized solar cell devices. Chapter 8, "Bifacial Photovoltaic Technology: Recent Advancements, Simulation and Performance Measurement", introduces the physic principle and applications of bifacial PV technology. This chapter presents different bifacial PV cell and module technologies as well as the advantages of using bifacial PV technology in the field. It discusses the advanced techniques for the characterization of bifacial PV modules and albedo as one of the important factors for the energy yield of bifacial PV technology. It also presents several simulation models and experimental measurements by varying the sensor positions on the rear side of the PV modules, different places, different albedo numbers, mounting heights, and different geographical locations with various tilts, seasons, and weather types. Chapter 9, "Photovoltaic Power Forecasting Methods", gives different physical, heuristic, statistical, and machine learning-based methods for PV power forecasting with several examples of their applications and related uncertainty. It also assesses the effect of degradation on lifetime PV energy forecast using linear and nonlinear degradation scenarios. Chapter 10, "Concentrator Photovoltaic System (CPV): Maximum Power Point Techniques (MPPT) Design and Performance", studies the performance of MPPT techniques applied to the CPV system for the research and the pursuit of the maximum power point (MPP). This chapter presents modeling and simulation of the CPV system including a PV module located in the focal area of a parabolic concentrator, a DC / DC converter (Boost), two MPPT controls (P&O and FL), and a resistive load. Finally, Chapter 11, "Model Reference Adaptive Control of Solar Photovoltaic Systems: Application to a Water Desalination System", deals with a new mathematic development of tracking control technique based on Variable Structure Model Reference Adaptive Following (VSMRAF) control applied to systems coupled with solar sources. This chapter provides a new theoretical analysis validated by simulation and experimental results to assure optimum operating conditions for solar PV systems with application in a water desalination system.

During my academic journey, I had the good fortune and privilege to be involved in numerous stimulating, provocative, and engaging classes, conversations, discussions, debates, workshops, seminars, and lectures on the energy sector at leading universities, industries, and institutions across the world. There are therefore so many people who have influenced my experience and expertise that led to numerous outcomes in the field of energy systems, renewables, PV solar energy, and integrated photovoltaics. This book is one of my latest publications, after two years of endeavor, and I hope it will be beneficial for readers ranging from energy industries, energy stakeholders, and energy policymakers to undergraduate and postgraduate students, young or experienced researchers, and engineers.

I would like to acknowledge all the authors who contributed to this book by proposing several interesting relevant topics. I am deeply indebted to colleagues, past and present, at Politecnico di Milano, Fraunhofer Institute for Solar Energy Systems (ISE), University of Freiburg, Helmholtz-Zentrum Berlin (HZB), Eindhoven University of Technology (TU/e), Amirkabir University of Technology (AUT), Universidade Federal de Santa Catarina (UFSC), and Norwegian University of Science and Technology (NTNU).

I want to express gratitude to my wife Shima and my lovely daughter Sana for their support and patience. My greatest debt is to my family, my parents (Naser and Horiyeh), my brother and sister (Ali and Zahra), my parents-in-law (Hussain and Masoumeh), and my sister- and brother-in-law (Mahsa and Amin).

I would like to offer special thanks to my uncle, Dr. Ebrahim Aghaei, who, although no longer with us, continues to inspire me with his great support and dedication in the past. He always believed in my ability to be successful in the academic arena ("You are gone but your belief in me has made this journey possible").

In the end, praise be to Allah who bestowed me with patience to accomplish editing this book. Without his mercy, this work could never have been done.

I would like to dedicate this humble work to Imam Muhammad b. al-Hasan al-Mahdi (a), who is the promised savior, the avenger of the blood of imam Hussain (a.s.), and who will rise one day and make the world full of peace and justice.

Mohammadreza Aghaei
Department of Ocean Operations and Civil Engineering,
Norwegian University of Science and Technology (NTNU),
Ålesund, Norway

Department of Sustainable Systems Engineering (INATECH),
University of Freiburg,
Freiburg, Germany

Section 1

Introduction

Chapter 1

Introductory Chapter: Solar Photovoltaic Energy

Mohammadreza Aghaei, Amir Nedaei, Aref Eskandari and Jafar Milimonfared

1. Introduction

The concept of energy transition is defined as a transformation of fossil-based energy resources to non-carbonated during the upcoming years [1]. Hence, supplying energy through renewable resources that can be naturally replenished on a human timescale is being of great importance. This form of energy is named renewable energy and is mostly sustainable and environmentally friendly. Renewable energy can be easily converted into different types of energy (e.g., electricity, heat) *via* recent technologies.

Accordingly, in 2015, the international community set the Sustainable Development Goals (SDGs) as a part of the UN 2030 Agenda for Sustainable Development [2]. The goals include pledges to eliminate poverty, starvation, etc. Of all the goals set by the international community, some goals such as to supply clean energy, to protect the climate, and so on were energy-related. It is mentioned that the seventh goal (known as SDG7) attempts to provide services for clean, affordable, and modern energy all over the world and increase the portion of renewable energy among the other types by 2030. Also, since all countries in the world are prone to suffer from climate change, SDG13 tries to increase the immunity by either enhancing the resilience of different countries or educating people and raising awareness.

2. Solar radiation

The main source of energy to move the atmosphere is the sun. This energy is radiated in the form of electromagnetic waves with a wavelength between 0.2 and 4 μm (see **Figure 1**). The smallest measurable amount of an electromagnetic field is called a photon. The modernized definition of photon is derived from research (which were based on those carried out by the German physicist Max Planck) done by Albert Einstein from 1900s to 1920s. In 1926, the term "photon" was popularized by Gilbert Lewis in his letter written to the Nature magazine.

The power that is received from the sun in the form of solar electromagnetic radiation per unit area over a given time period is named solar irradiance and is measured in W/m^2 in SI units. Irradiance can be measured in space or at the Earth's surface after it has partially been absorbed by the atmosphere as well as scattered. On the Earth's surface, the amount of irradiance is a function of the tilt of the measuring surface, the height of the sun above the horizon, and also the atmospheric conditions. **Figure 2** depicts the irradiance of the sun at the Earth's surface in both a direct normal irradiation (DNI) and a global horizontal irradiation (GHI).

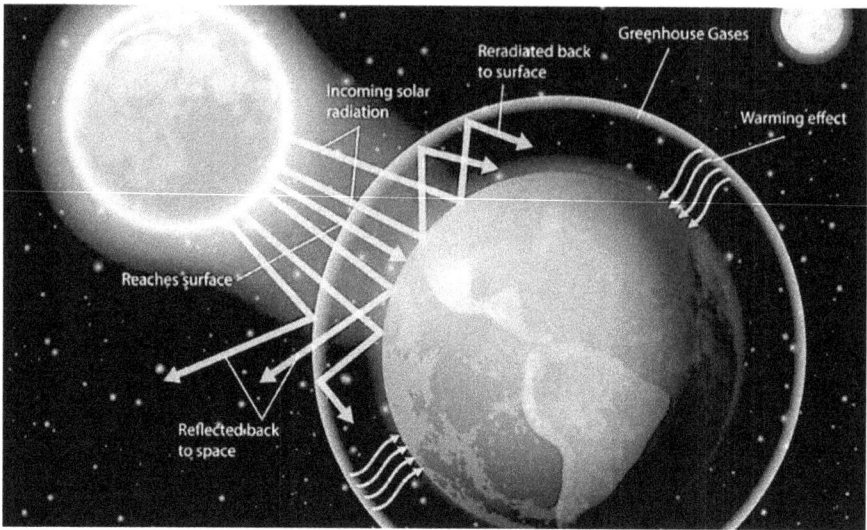

Figure 1.
The Earth's radiation budget [3].

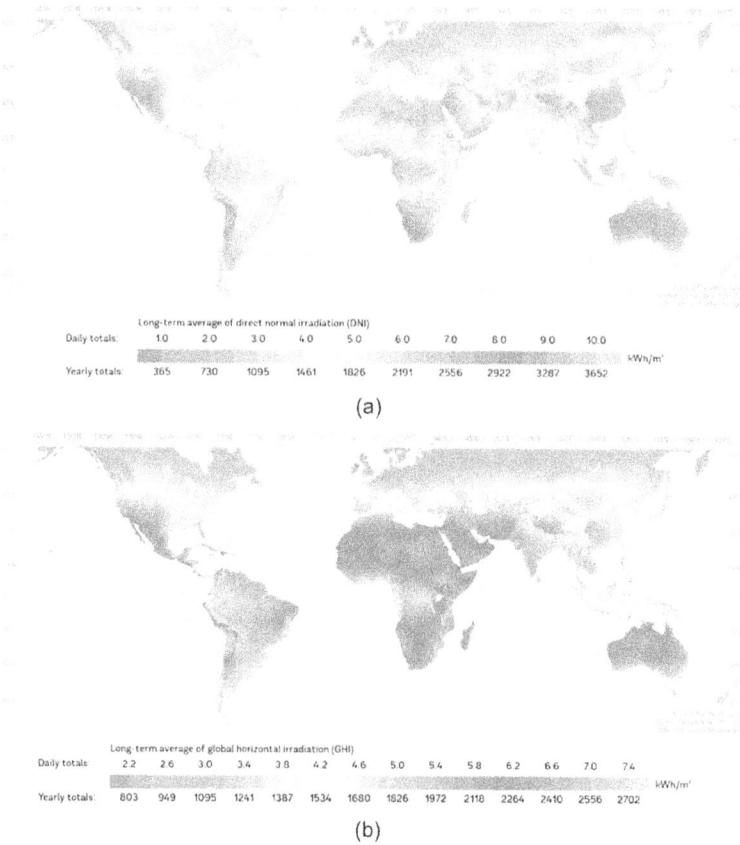

(a)

(b)

Figure 2.
(a) The world's map of the direct normal solar irradiation (DNI) at the Earth's surface and (b) the world's map of the global horizontal irradiation (GHI) at the Earth's surface according to Global Solar Atlas 2.0 [4].

3. Solar energy

The light and heat that are radiated from the sun are often named solar energy and are one of the most significant sources of renewable energy. Solar energy can be harnessed through some technologies that are categorized into two main classes namely active solar technologies such as photovoltaic systems and passive solar technologies that include a wide variety of techniques such as orienting a building to the sun.

3.1 Solar photovoltaics

The history of photovoltaics (PV) dates back to 1800s when Alexandre Edmond Becquerel observed PV effect. This was followed by testing the first solar cell with the efficiency of less than 1% in 1883. It was then in the first two decades of the twentieth century when Albert Einstein published his paper on photoelectric effect that resulted in his first and only Noble Prize in 1921. A decade later, in 1931, the first pure semiconductor was developed. At first, in 1950s, solar cells were utilized for space applications. In 1957, solar cells with around 8% of efficiency were developed, a record that was soon broken by Hoffman Electronics to a high of 10 and 14% in 1959 and 1960, respectively. Soon after, the first amorphous silicon PV cell was developed and the global PV capacity rose to 500 kW. This amount grew even further and reached a high of 21.3 MW in 1983. In about 20 years, in 2002, 175-kW high-concentrating PV plant was installed in Arizona, United States. Four years later, the world witnessed a new record of 40% efficiency for PV technology. With the increase in the global PV capacity to 100 GW in 2012, the manufacturing costs reduced significantly to $1.25 per watt. In 2016, the first solar-powered plane flew around the world [5]. **Figure 3** depicts the PV power potential in the world.

Solar photovoltaic generation has broken the record of 156 GWh (23%) in 2020 to reach 821 GWh, which proved the second largest growth of all renewable technologies in 2020, slightly behind wind and ahead of hydropower. In China, the United States, and Vietnam, an unprecedented surge (a record of 134 GW) in PV capacity additions took place. Solar PV is undeniably becoming the lowest cost option for electricity generation all around the world and is expected to attract a vast amount of investment in the coming years [6].

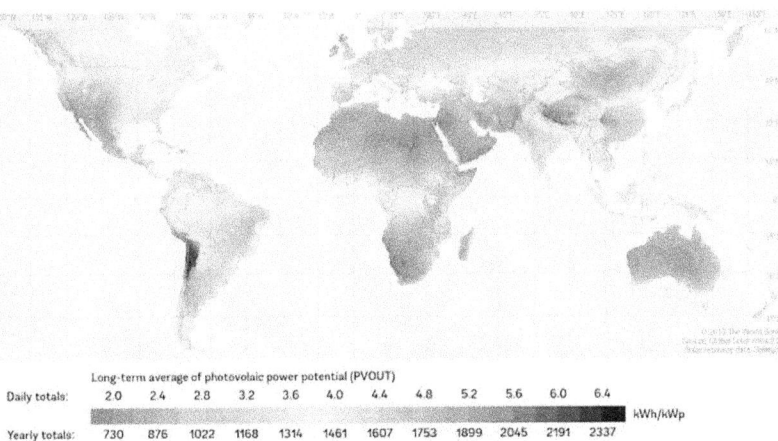

Figure 3.
The world's map of photovoltaic power potential [4].

Furthermore, solar energy is predicted to play a key role in the future global energy system owing to the scale of the solar resource. The installed solar photovoltaic (PV) throughout the world exceeded 1 TW at the beginning of 2022. This brought the world into the era of TW-scale PV [7]. This will definitely be fortified by the rapid expansion of PV industry as well as everyday cost decreases. The world envisions a future with nearly 10 TW of PV by 2030 and 30–70 TW by 2050, which can provide a majority of global energy [8].

3.2 Solar cells

The term "solar cell" was previously mentioned in the history of photovoltaics. In fact, solar cell is attributed to any device that directly converts the energy of light into electrical energy through the photovoltaic effect. The vast majority of solar cells are fabricated from silicon with rising efficiency and decreasing cost as the materials range from amorphous (non-crystalline) to polycrystalline and mono-crystalline (single crystal) silicon forms. Solar cells, in comparison with batteries or fuel cells, do not utilize chemical reactions or require fuel to produce electricity, and, compared with electric generators, they do not have any moving parts [9].

As previously mentioned, solar cells are usually categorized into four main classes including the following:

(1) Monocrystalline solar cells that are also known as single crystalline cells and are very easy to identify due to their dark black color. They are made from a very pure form of silicon that has made them become the most efficient material for the process of sunlight conversion into electricity, (2) polycrystalline cells (multi-silicon cells) that were the first solar cells to be developed in the industry in the beginning of 1980s, (3) amorphous solar cells that, as the word "amorphous" meaning "shapeless" suggests, are not structured or crystallized on a molecular level and were commonly used for small-scale applications, and (4) thin film solar cells that are manufactured by placing several thin layers of photovoltaic on top of each other to create the module [10].

Another important concept in this area is named "the spectral response." A solar cell's spectral response to light of a single wavelength is its response at that specific wavelength multiplied by the intensity of the light. If the actual irradiance and device spectral response profiles are symmetrical around the center wavelength, then the currents generated from light on each side of the center are equal, and their

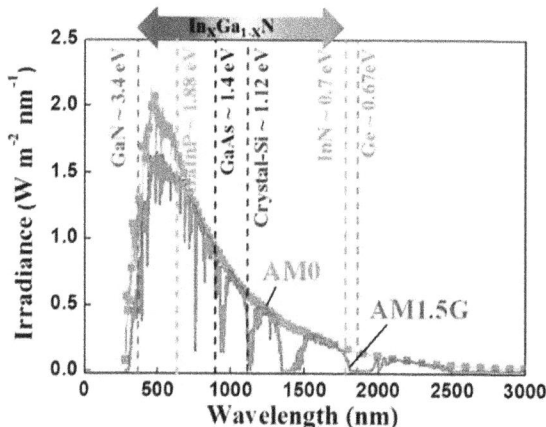

Figure 4.
The solar spectral irradiance at air mass 0 (AM0) and global air mass 1.5 (AM1.5G) and the cutoff wavelength of semiconductor materials for common solar cell applications [11].

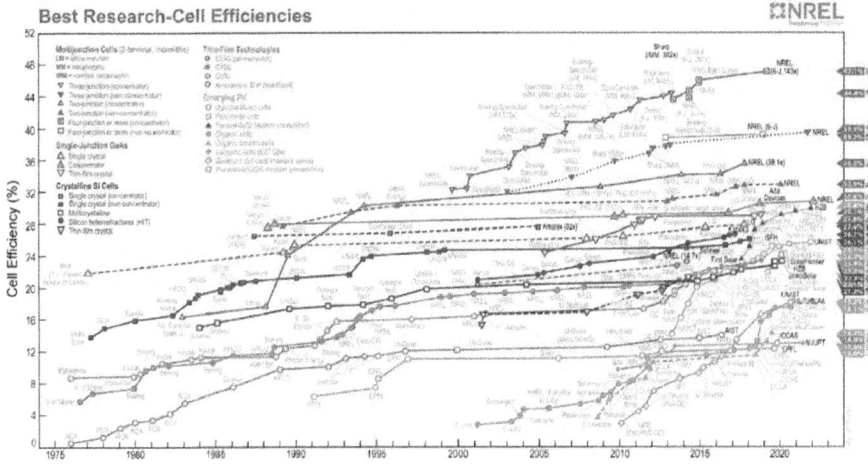

Figure 5.
Solar cell efficiencies throughout the history [13].

sum is equivalent to the current that the device would generate if illuminated by a single-wavelength source of the same intensity (see **Figure 4**) [12].

On the other hand, efficiency is the most commonly used parameter to compare the performance of one solar cell to another. It is defined as the ratio of the output energy from a solar cell to the input energy from the sun. Moreover, the efficiency also depends on the spectrum and intensity of the sunlight and the temperature of the solar cell. Therefore, conditions under which efficiency is measured must be carefully controlled in order to properly compare the performance of one device to another. Terrestrial solar cells are measured under AM1.5 conditions and at an ambient temperature of 25°C. Whereas for space uses, solar cells are measured under AM0 conditions. A comprehensive report on the past, and recent and projected solar cell efficiency results is provided in **Figure 5** [13].

3.3 Solar photovoltaic systems

Solar cells are arranged into large groupings, which are called solar arrays. These arrays, composed of thousands of solar cells, can be considered as central electric

Figure 6.
An PV power plant located in Hungary [14].

power stations which convert sunlight into electrical energy to be distributed to industrial, commercial, and residential consumption. On the other hand, in a smaller scale, the configuration is commonly referred to as solar modules, which are mostly installed by homeowners on their rooftops to replace their conventional electric supply. Solar modules are also used to provide electric power in many remote areas where conventional electric power sources are either unavailable or prohibitively expensive to install. Solar cells also provide power for most space installs, from communications and weather satellites to space stations owing to the fact that they do not have any moving parts; therefore, there is no need for maintenance or any fuels that would require replenishment. Solar cells (as will be discussed further ahead) have also been used in consumer products, such as electronic toys, calculators, and radios [9]. However, in a large-scale version, in solar PV plants (see **Figure 6**), thermal energy from the sun is utilized and further transformed into electrical energy using photovoltaic modules installed in an optimal configuration. The thermal energy is abundant, easy to access, and cheap. Another type of solar power plant (which does not seem to be as common as the previous type) is the concentrated solar power plant, which contains plenty of mirrors or lenses that are carefully placed in an organized way to concentrate on collected heat to one specific position, which is further utilized to supply power for a steam turbine that generates electricity [15].

3.4 Current global status of photovoltaics

According to Feldman et al. [16], from 2010 to 2020, the addition to global PV capacity grew from 17 to 139 GW_{DC} in a way that the global PV installations reached 760 GW_{DC} at the end of 2020. In 2020, approximately 100 MW of concentrated solar power was added in China. At the end of that year, 57% of cumulative PV installations were in Asia, 22% were in Europe, and 15% were in the Americas. The United States is now the country with the second largest cumulative installed PV capacity.

Also, China, the United States, Japan, Germany, and India were the leading five markets in cumulative PV installations at the end of 2020. However, Vietnam, with more than 11 GW of installations in 2020, took India away from the top five for annual deployment.

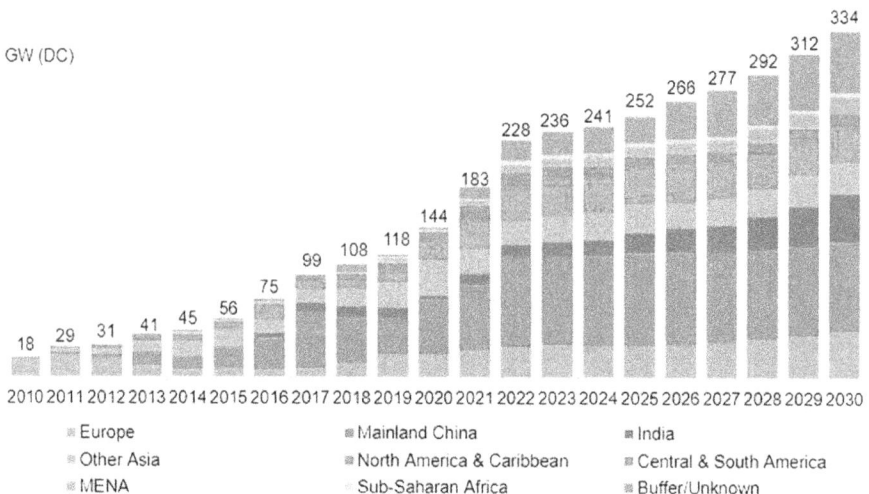

Figure 7.
Current and projected installations throughout the world [17].

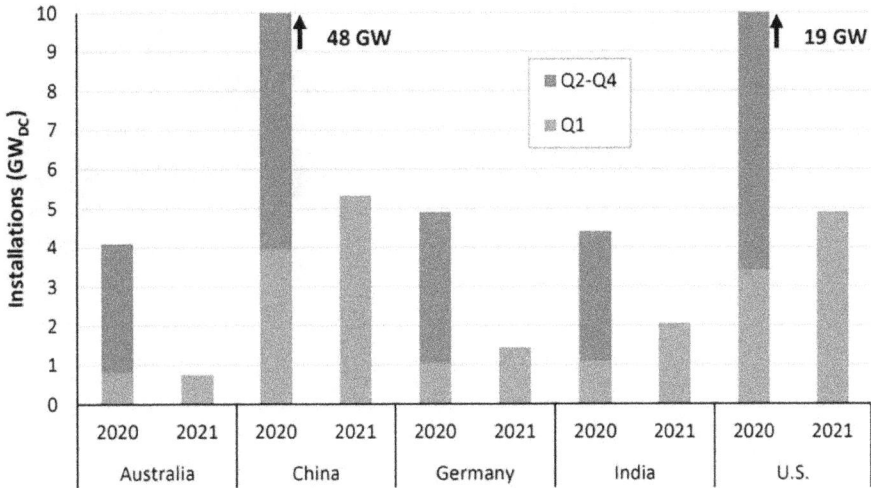

Figure 8.
Installation growth from 2020 to 2021 [18].

According to IEA estimation, in 2020, PV was the main source of 3.7% of global electricity generation. Although the United States was a leading PV market, it was below the average and other leading markets concerning PV generation as a percentage of total country electricity generation, with 3.4%.

	Location	Capacity (MWp)	Area (km²)	Year of installation	Photograph
1	Gonghe County, China	2200	50	2021	
2	Abu Dhabi, the UAE	1177	8	2019	

	Location	Capacity (MWp)	Area (km²)	Year of installation	Photograph
3	Yanchi, China	820	—	2016	
4	Datong, China	800	—	—	
5	Zaragoza, Spain	730	31.7	2020	
6	Villanueva, Mexico	828	27.5		

	Location	Capacity (MWp)	Area (km²)	Year of installation	Photograph
7	Nevada's Eldorado Valley, USA	816	16.2	2010	
8	California, USA	747	13	—	
9	Kamuthi, India	648	10.1	2016	
10	Jaisalmer, India	600	—	—	

	Location	Capacity (MWp)	Area (km²)	Year of installation	Photograph
11	Hongshagang, China	574	—	—	
12	Topaz, USA	550	19	2014	
13	Sao Goncalo, Brazil	549	13	—	
14	Yinchuan, China	500	—	2018	

Table 1.
The world largest solar power plants [19].

From Q1 2020 to Q1 2021, installations in China, the United States, and Germany increased from 35 to 45%, and specifically those in India rose 89% although analysts argued that India's large increase was due to developers finishing delayed 2020 projects. Despite the growth in installations, it was not necessarily indicative of 2021 as a whole. A significant portion of deployment often comes toward the end of the year. Significant supply constraints, increased costs, and resurgent waves of the pandemic (particularly in India) might suppress installations, see **Figure 7**.

Analysts also predict continued growth in annual global PV installations, with a median estimate of 209 GW_{DC} in 2022 and 231 GW_{DC} in 2023. China, Europe, the United States, and India are anticipated to involve in about two-thirds of global PV installations over this period. Analysts note that these projections come despite many projects in 2022 risking delay or cancelation because of increasing material and shipping costs, see **Figure 8** [11]. **Table 1** lists the top 14 PV power plants around the world.

4. Trends and applications of photovoltaics

Photovoltaic technology has many applications to improve human life. To date, many applications of this technology have been utilized in industry as well as by ordinary users (see **Figure 9**). However, the advances in this technology do not certainly end with its current applications, and therefore, it sees many bright horizons ahead. Numerous examples of new applications of photovoltaic technology are as follows.

4.1 Building-attached photovoltaic (BAPV)

BAPV is the classic arrangement of photovoltaic systems and solar cells mounted on the roofs or building surfaces. Although it is probable to be aesthetically problematic, this is to avoid any sort of shading as much as possible. Moreover, in this

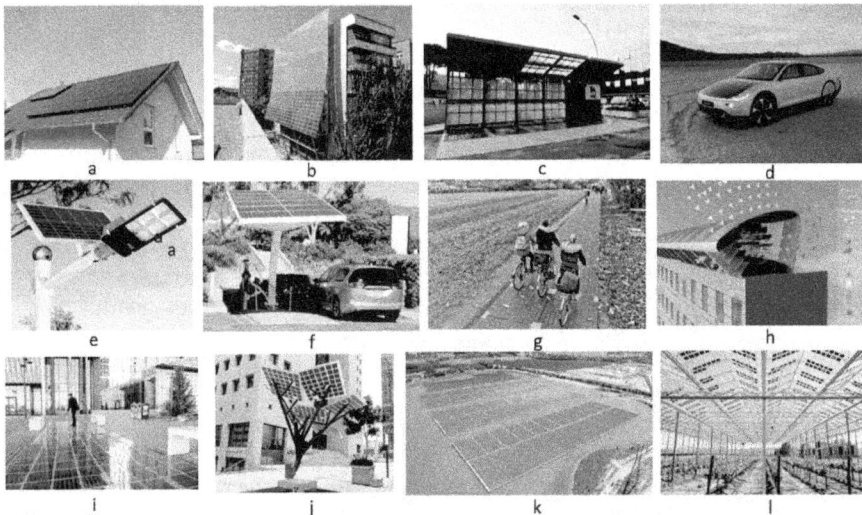

Figure 9.
Various examples of novel applications of photovoltaic technology are as the follows: (a) BAPV [20], (b) BIPV [21], (c) LSC PV [22], (d) VIPV [23], (e) solar street lights [24], (f) PV charging stations [25], (g) PV bike path [26], (h) PV windrail [27], (i) solar-powered pavement [28], (j) PV tree [29], (k) floating PV systems [30], and (l) agri-voltaic system [31].

application, the installer is required to utilize several pieces of equipment such as those needed for the mounting system.

4.2 Building-integrated photovoltaic (BIPV)

In comparison with BAPV, BIPV is an innovative design in which solar cell and in general the photovoltaic system are integrated with the construction itself, either the façade (e.g., walls, windows) or the roofs as skylights, shingles, etc. [32]. Therefore, it is targeted to be practiced as a phase of aesthetics with any utilitarian views.

4.3 Luminescent solar concentrator photovoltaic (LSC PV)

LSC PV devices consist of transparent optical light guides typically made of a polymer or glass containing luminophores with one or more photovoltaic (PV) solar cells mounted on one or more edges and sometimes rear of the light guide [33, 34]. The sunlight is then intercepted by small photovoltaic cells and sequently converted into electricity. The main advantage of LSC is that it is capable of producing electricity even in low-light conditions and can be incorporated into architectural structures particularly as transparent elements. LSC's outstanding versatility is undeniable since it can be integrated either with houses and buildings as a colored window, a leaf roof, a smart window, etc., or with urban facilities as a noise barrier, a parking shed, etc.

Recently, Aghaei et al. have developed attractive mosaic LSC PV devices made by miniaturizing cubical light guides and mounting bifacial solar cells to the edges of neighboring light guides, as well as optionally attaching monofacial PV to the bottom sides. These mosaic LSC PV devices could be applied to make solar energy ubiquitous to the urban setting where it requires making visually appealing devices that can function in the challenging lighting conditions found in cities. Thus, by developing such colorful, visually appealing mosaic LSC PV devices, one can accelerate the general acceptance of solar energy in the built environment, even with the modest efficiency devices [35].

4.4 Vehicle-integrated photovoltaic (VIPV)

Although photovoltaic technology is mostly utilized in grid-connected applications, a new application could be the integration of photovoltaic into battery electric vehicles, creating a VIPV. Photovoltaic can help recharging the vehicle battery, without being connected to a charging station. VIPV can make transport more sustainable and seems to be cost-efficient. Accompanied with a systematic appropriate installation, it can also appear in solar cars, buses, spacecrafts, boats, UAVs (unmanned aerial vehicles), trains, hybrid airships, AUVs (autonomous undersea vehicles), bikes, etc.

4.5 Solar street lights

Another impactful application of solar cells is to be installed especially on top of street or roadway lights as a power supply. The installment is usually accompanied with an oversized battery not only to power the system at nights or in low-light conditions but also to enable an autonomous performance in aforementioned conditions up to even five days [36]. Furthermore, the autonomy helps the system to remain needless to be constantly connected to the grid so that the whole system will be able to work properly for a specific period of time even off-grid.

4.6 Photovoltaic charging stations

Slightly different from VIPVs, charging stations are usually roof-mounted and much simpler in terms of power electronic devices. These stations, therefore, are lighter in weight and smaller in the area they occupy. As a result, the stations are often cheaper and easier to be maintained for a long time.

4.7 Photovoltaic bike path

A tremendous amount of sunlight scatters regularly all over the ground, which can be harvested with the aid of specifically designed machines, vehicles, and equipment. An example of these vehicles is a bike ridden along a solar-cell-covered path and is known as a PV bike path. The bike is designed in a way that the wheels are in a high level of friction with the path [37]. This might be the point of entry into the maximization of land utilization along with power generation.

4.8 Photovoltaic-integrated zero/low-energy buildings

Zero-energy buildings (ZEBs) generally refer to a category of buildings with very high energy performance, characterized by a very low or approximately zero annual energy requirement. The required amount of energy is entirely or significantly covered by renewable energy, including energy from renewable sources produced on-site or nearby [38]. This renewable energy can be comprised of PV systems being used to supply electric energy demand, etc.

4.9 Photovoltaic-powered air conditioners

Air conditioning has always been an important issue either in industrial, commercial, or residual consumption. The use of PV systems in this area can produce a noticeable reduction in energy costs and bring about economic benefits. Also, using a PV-powered air conditioner has proved to conserve nearly 67 and 77% of the grid energy in summer during the day and at night, respectively [39].

4.10 Photovoltaic windrail

Owing to the great flexibility of PV technology, PV windrails are a novel combination of both wind and solar energy. The device is mostly installed on the rooftop of a building to obtain the maximum speed of the blowing wind. The wind then heads toward a channel to generate electricity using turbine generators.

4.11 Photovoltaic trees

PV trees are one of the most intriguing applications, which can be achieved from PV systems. The problem of land shortage or even the urban aesthetics can cause the system to be lifted and mounted on top of steel stems, which physically resembles a tree with PV modules as the leaves on top. The system can bring numerous advantages from simply charging small handy gadgets to supplying the power needed for street lights and electric vehicles. The future of PV trees seems to be extremely promising.

4.12 Photovoltaic pavements

Pavements have covered 30–40% of the urban surface [40] and are considered as enormous potential for PV installments and energy generation through solar PV modules. Surprisingly, studies have shown that even the temperature on the surface of walkable PV pavements or cycling tracks is proved to be lower than that on conventional pavements.

4.13 Landscape-integrated photovoltaic

A novel idea in PV systems installation is to integrate the natural landscape with solar PV modules. The idea is developed in a way that bifacial PV modules can be utilized vertically to avoid using flat spacious surfaces, rooftops, etc., and therefore taking advantage of more installation space to increase the amount of power generated by the system.

4.14 Product-integrated photovoltaic (PIPV)

Integrated with several products, solar photovoltaic energy can be exploited to yield grid-independent, battery saving, and wasteless devices. Many calculators and watches have been utilizing solar energy for ages to supply power. Moreover, the area of usage can be (and even is in some cases) expanded to lamps, chargers, scales, etc., even though the design is definitely under development and progression for further utilizations [41].

4.15 Floating photovoltaic systems

In areas suffering from land limitation, the concept of floating photovoltaic systems and utilizing the area on the surface of water can be highly advantageous. The system is even capable of being integrated with hydropower electricity transmission system to end up a higher efficiency [30].

4.16 Submerged photovoltaic

Another novel idea that can be achieved through the flexibility of PV technology is to use the spacious area underwater to locate the PV arrays and to take advantage of the natural cooling system. One appropriate space is a swimming pool where the modules can be installed both on the edges and on the pool floor.

4.17 Agri-voltaic system

The term "Agri-voltaic system" refers to a combined production of photovoltaic power and agricultural products on the same area of land. Solar modules and crops share light and radiation so that modules that are located above part of the crops generate shade and create a kind of microclimate over the mentioned area. Therefore, the result will be more fresh products, less water requirement for the plants, and lower losses due to evaporation [42].

5. Summary

To wrap up, the world has long been transitioning from fossil fuels to renewables. Not only does this help preserve the environment, but it also brings about other benefits under the sustainable development goals (SDGs) by the international community, including the fight against poverty and hunger. Moreover, photovoltaic technology has been facilitating people's lives for many years. Over 183 GW of photovoltaic systems was installed worldwide, which is nearly 40 GW more than

2020. Many countries, including China, the United States, and India, have paid special attention to photovoltaic technology to meet their electrical and thermal needs, both in industry and in the field of home consumption. As of 2020, cumulative installed solar power capacity in China that leads the whole world in this field had reached almost 253 GW. Also, as mentioned, with the increase in progress in the photovoltaic industry and also the daily increasing reduction of prices in this field, the solar resources will reach the level of Terawatt-scale in the coming years. Moreover, areas of photovoltaic use are transitioning from conventional to more advanced areas such as PV pavements, BIPV, agri-voltaic systems. Undoubtedly, due to the increasing advances in photovoltaic technology, its use in the future will be much wider and more common than today.

Author details

Mohammadreza Aghaei[1,2*], Amir Nedaei[3], Aref Eskandari[3] and Jafar Milimonfared[3]

1 Department of Ocean Operations and Civil Engineering, Norwegian University of Science and Technology (NTNU), Alesund, Norway

2 Department of Sustainable Systems Engineering (INATECH), University of Freiburg, Freiburg, Germany

3 Department of Electrical Engineering, Amirkabir University of Technology, Tehran, Iran

*Address all correspondence to: mohammadreza.aghaei@ntnu.no

IntechOpen

References

[1] International Renewable Energy Agency. Renewable Capacity Statistics 2022 [Statistiques De Capacité Renouvelable 2022 ESTADÍSTICAS De Capacidad Renovable 2022]. 2022 [Accessed: 03 May 2022]

[2] Communications Materials—United Nations Sustainable Development. Available from: https://www.un.org/sustainabledevelopment/news/communications-material/. [Accessed: 03 June 2022]

[3] World Map/World Atlas/Atlas of the World Including Geography Facts and Flags—WorldAtlas.com. Available from: https://www.worldatlas.com/. [Accessed: 31 May 2022]

[4] Global Solar Atlas 2.0. A free, web-based application is developed and operated by the company Solargis s.r.o. on behalf of the World Bank Group, utilizing Solargis data, with funding provided by the Energy Sector Management Assistance Program (ESMAP). Available from: https://documents1.worldbank.org/curated/en/529431592893043403/pdf/Global-Solar-Atlas-2-0-Technical-Report.pdf. [Accessed: 03 May 2022]

[5] https://www.smithsonianmag.com/innovation/inside-first-solar-powered-flight-around-world-180968000/

[6] Solar PV Power Generation in the Net Zero Scenario, 2000-2030—Charts—Data and Statistics—IEA. Available from: https://www.iea.org/data-and-statistics/charts/solar-pv-power-generation-in-the-net-zero-scenario-2000-2030. [Accessed: 03 May 2022]

[7] https://www.pv-magazine.com/2022/03/15/humans-have-installed-1-terawatt-of-solar-capacity/

[8] Haegel NM et al. Terawatt-scale photovoltaics: Transform global energy improving costs and scale reflect looming opportunities. Science (1979). 2019;**364**(6443):836-838

[9] Solar Cell | Definition, Working Principle, & Development | Britannica. Available from: https://www.britannica.com/technology/solar-cell. [Accessed: 01 June 2022]

[10] The Renewable Energy Hub. Available from: https://www.renewableenergyhub.co.uk/. [Accessed: 25 May 2022]

[11] Yang C-C et al. Characteristics of InGaN/sapphire-based photovoltaic devices with different superlattice absorption layers and buffer layers. Gallium Nitride Materials and Devices VI. 2011;**7939**:269-275

[12] Field H, Field H. Solar cell spectral response measurement errors related to spectral band width and chopped light waveform. In: Conference Record of the Twenty Sixth IEEE Photovoltaic Specialists Conference - 1997. 1997. Available from: https://ieeexplore.ieee.org/document/654130

[13] Best Research-Cell Efficiency Chart | Photovoltaic Research | NREL. Available from: https://www.nrel.gov/pv/cell-efficiency.html. [Accessed: 01 May 2022]

[14] Hungary's Largest Solar Power Plant is Now Operational | Sun & Wind Energy. Available from: https://www.sunwindenergy.com/photovoltaics/hungarys-largest-solar-power-plant-now-operational. [Accessed: 03 June 2022]

[15] Belyakov N. Energy system and basic electricity market. Sustainable Power Generation. 2019:63-87. DOI: 10.1016/B978-0-12-817012-0.00012-8

[16] Feldman D, Wu K, Margolis R. H1 2021 Solar Industry Update, 2021

[17] BloombergNEF Says Global Solar Will Cross 200 GW Mark for First Time This Year, Expects Lower Panel Prices: Neoliberal. 2022. Available from: https://www.reddit.com/r/neoliberal/comments/si51w1/bloombergnef_says_global_solar_will_cross_200_gw/. [Accessed: 01 May 2022]

[18] Feldman D, Dummit K, Zuboy J, Heeter J, Xu K, Margolis R. Winter 2021/2022 Solar Industry Update, 2022

[19] https://www.pv-magazine.com/2021/09/09/the-worlds-largest-solar-power-plants/

[20] Types of Solar Panels. Available from: https://www.the-green-house.net/types-of-solar-panels/. [Accessed: 01 June 2022]

[21] Building Integrated Photovoltiacs (BiPV) Can Be Low Cost. Available from: https://www.kschan.com/building-integrated-photovoltaics-bipv-can-be-low-cost/. [Accessed: 01 June 2022]

[22] Luminescent Solar Concentrator LSC | EnergyCuE. Available from: https://energycue.it/luminescent-solar-concentrator-lsc/9655/. [Accessed: 01 June 2022]

[23] https://lightyear.one/

[24] 400W LED Solar Street Lights Outdoor, Dusk to Dawn Security Flood Light with Remote Control & Pole, Wireless, Waterproof, Perfect for Yard, Parking Lot, Street, Garden and Garage. Amazon.com. Available from: https://www.amazon.com/Outdoor-Security-Control-Wireless-Waterproof/dp/B07Z4T5KPW?th=1. [Accessed: 01 June 2022]

[25] Goleta Council Approves Solar Photovoltaic Project, EV Charging Stations for City Hall | Local News—Noozhawk.com. Available from: https://www.noozhawk.com/article/goleta_city_council_approves_solar_photovoltaic_project_ev_charging_station. [Accessed: 01 June 2022]

[26] The First Solar Cycle Lane Opens in Germany, to Harness the Power of the sun—LifeGate. Available from: https://www.lifegate.com/germany-first-solar-cycle-lane. [Accessed: 01 June 2022]

[27] An Urban Wind and Solar Energy System That May Actually Work. Available from: https://www.bdcnetwork.com/urban-wind-and-solar-energy-system-may-actually-work. [Accessed: 01 June 2022]

[28] Platio Integrates Solar Panels in Pedestrian Walkways and Street Furniture. Available from: https://newatlas.com/platio-photovoltaic-solar-panel-paving/50549/. [Accessed: 01 June 2022]

[29] What Are Solar Trees? How Do They Compare to Solar Panels? Available from: https://www.treehugger.com/what-are-solar-trees-5207944. [Accessed: 01 June 2022]

[30] Europe's Largest Floating Solar System to be Powered by 17MW of PV Modules. Available from: https://energyindustryreview.com/renewables/europes-largest-floating-solar-system-to-be-powered-by-17mw-of-pv-modules/. [Accessed: 03 June 2022]

[31] To Feed a Growing Population, Farmers Look to the Sun | Research and Innovation. Available from: https://ec.europa.eu/research-and-innovation/en/horizon-magazine/feed-growing-population-farmers-look-sun. [Accessed: 01 June 2022]

[32] Yoon JH, Song J, Lee SJ. Practical application of building integrated photovoltaic (BIPV) system using transparent amorphous silicon thin-film PV module. Solar Energy. 2011;85(5):723-733

[33] Aghaei M, Zhu X, Debije M, Wong W, Schmidt T, Reinders A. Simulations of luminescent solar concentrator bifacial photovoltaic mosaic devices containing four different organic luminophores. IEEE Journal of Photovoltaics. 2022;**12**(3):771-777

[34] Rafiee M, Chandra S, Ahmed H, McCormack SJ. An overview of various configurations of luminescent solar concentrators for photovoltaic applications. Optical Materials (Amst). 2019;**91**:212-227

[35] Aghaei M, Pelosi R, Wong WWH, Schmidt T, Debije MG, Reinders AHME. Measured power conversion efficiencies of bifacial luminescent solar concentrator photovoltaic devices of the mosaic series. Progress in Photovoltaics: Research and Applications. 2022;**30**(7):726-739

[36] Vieira JAB, Mota AM. Implementation of a stand-alone photovoltaic lighting system with MPPT battery charging and LED current control. In: 2010 IEEE International Conference on Control Applications. 2010. pp. 185-190. DOI: 10.1109/CCA.2010.5611257. Available from: https://www.infona.pl/resource/bwmeta1.element.ieee-art-000005611257

[37] Shekhar A et al. Harvesting roadway solar energy-performance of the installed infrastructure integrated PV bike path. IEEE Journal of Photovoltaics. 2018;**8**(4):1066-1073

[38] Bisegna F, Evangelisti L, Gori P, Guattari C, Mattoni B. From efficient to sustainable and zero energy consumption buildings: Green buildings rating systems. In: Handbook of Energy Efficiency in Buildings: A Life Cycle Approach. Applied Sciences, MDPI; 2018. pp. 75-205. Available from: https://www.mdpi.com/2076-3417/12/4/2136/htm

[39] Liu Z, Li A, Wang Q, Chi Y, Zhang L. Performance study of a quasi

grid-connected photovoltaic powered DC air conditioner in a hot summer zone. Applied Thermal Engineering. 2017;**121**:1102-1110

[40] (PDF) Characterizing the Fabric of the Urban Environment: A Case Study of Metropolitan Chicago, Illinois 1. Available from: https://www.researchgate.net/publication/280091683_Characterizing_the_Fabric_of_the_Urban_Environment_A_Case_Study_of_Metropolitan_Chicago_Illinois_1. [Accessed: 29 April 2022]

[41] CASO Body Solar Digital Bathroom Scales Weight Range=150 kg Silver | Conrad.com. Available from: https://www.conrad.com/p/caso-body-solar-digital-bathroom-scales-weight-range150-kg-silver-1497211. [Accessed: 01 June 2022]

[42] Agrivoltaic Systems: A Promising Experience. Available from: https://energyindustryreview.com/analysis/agrivoltaic-systems-a-promising-experience/. [Accessed: 01 May 2022]

Section 2

Fundamentals, Measurements and Modeling of Solar Radiation

Chapter 2

Measuring Solar Irradiance for Photovoltaics

Marc A.N. Korevaar

Abstract

In recent years, solar energy technology has emerged as one of the leading renewable energy technologies currently available. Solar energy is enabled by the solar irradiance reaching the earth. Here we describe the characteristics of solar irradiance as well as the sources of variation. The different components of the solar irradiance and the instruments for measurement of these components are presented. In photovoltaics, the measurement of solar irradiance components is essential for research, quality control, feasibility studies, investment decisions, plant monitoring of the performance ratio, site comparison, and as input for short-term irradiance forecasting. Some more details are also provided related to physics of measuring instruments, their calibration, and associated uncertainty.

Keywords: OCIS codes: (350.6050) solar energy, (010.5630) radiometry, (120.0120) instrumentation, measurement, and metrology

1. Introduction

1.1 The sun

The sun provides 99.97% of the energy at the earth's surface (the rest is geothermal), and it is responsible, directly or indirectly, for the existence of life on earth. The energy, generated by nuclear fusion of hydrogen, emitted by the sun, is approximately 63 MW for every m^2 of its surface, about 3.72×10^{20} MW in total. The surface of the sun is very hot, and the layer emitting most of the radiation, the photosphere, is at about 5770 Kelvin. This means that there is a lot of short-wave radiation, ultraviolet and visible, and it takes approximately 8.3 minutes to reach the earth.

The unit for the measurement of irradiance (radiative flux [1]) is watts per square meter (W/m^2). At the mean distance between the earth and sun of 150 million kilometers (1 astronomical unit (AU)), the total solar irradiance (TSI) reaching the Earth's atmosphere is $1,360.8 \pm 0.5$ W/m^2 at a solar minimum [2] (over all wavelengths and perpendicularly). This quantity is named the "Solar Constant" [3]. However, it is not actually constant as is shown in **Figure 1**. The solar activity [3] varies with an 11-year cycle by $\pm$ 0.1% [2, 5]. This variation coincides with the number of sunspots. More sunspots mean more solar activity (solar flares and coronal mass ejection) and therefore a slightly larger amount of solar irradiance reaching the earth. Additionally, there is a larger variation of the solar irradiance reaching the top of the atmosphere by the sun to earth distance variation. This

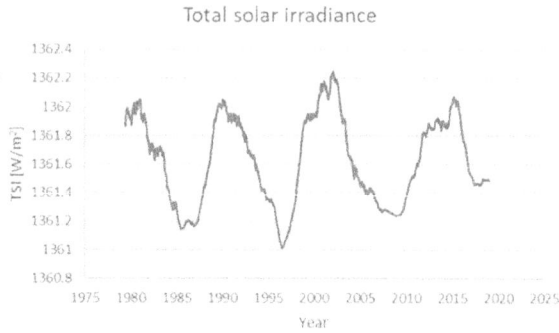

Figure 1.
Total solar irradiance (integrated over all wavelengths), or alternatively called "solar constant" [3], data over the last 45 years [4].

variation is due to the earth orbit eccentricity. At the perihelion (in January), the earth is close to the sun (147 million km), and at the aphelion (in July), the earth is further from the sun (152 million km).

This results in a variation of ± 3% in solar irradiance, described in [5], which can be approximated with the following equation [6]:

$$I(n) = I_0 \left[1 + 0.034 \cdot \cos \left(\frac{2\pi \cdot n}{365.25} \right) \right]$$
(1)

Where n is the day of the year, and I_0 is the solar constant. Interestingly enough the temperature is actually higher during the aphelion when the sun is further away. This is due to the tilt and land distribution; the land is distributed more in the northern hemisphere. During the northern summer, the North Pole is tiled more toward the sun, and more land is irradiated. The land heats up easily compared with oceans, and this leads to a higher temperature on earth when the earth is further away from the sun. When passing through the atmosphere, some solar radiation reaches the earth's surface as a direct beam, and some is scattered or absorbed by the atmosphere [5], aerosols (fine solid particles and liquid droplets), and clouds. Gaseous molecules, aerosols, and clouds cause most of the absorption, which heats up the atmosphere. All of the UV-C and most of the UV-B are absorbed by oxygen and ozone in the stratosphere.

1.2 Scattering

The amount of scattering of light is influenced by the length the light travels through the atmosphere. This length can be defined as the air mass:

$$AM = \frac{L}{L_0} \approx \frac{1}{\cos(\theta_z)}$$
(2)

Where L is the length through the atmosphere, Lo the length at perpendicular incidence, and θz the solar zenith angle [7].

The scattering processes can be described by two different processes: Mie scattering and Rayleigh scattering. The scattering process is determined by the particle diameter in relation to the wavelength of the light. For water droplets and ice crystals, the particle diameter is typically of the same order as the wavelength, and therefore, Mie scattering applies. Mie scattering is equal for all wavelengths and tends to give the scattering particles a white appearance, such as for clouds. An example of this phenomenon in nature is shown in **Figure 2a**.

For gas molecules in the atmosphere, the particle diameter is much smaller than the solar wavelength, and therefore, Rayleigh scattering applies. The scattering cross section for Rayleigh scattering is given by:

$$\sigma_s = \frac{2\pi^5}{3}\frac{d^6}{\lambda^4}\left(\frac{n^2-1}{n^2+2}\right)^2 \tag{3}$$

Where n is the particle refractive index, d the particle diameter, λ the light wavelength. The scattered fraction is given by:

$$\sigma_s \cdot N \tag{4}$$

Where N is the number of particles. What we can conclude from Eqs. (3) and (4) is that the shorter the wavelength is, the more scattering occurs. Additionally, a longer path through the atmosphere (air mass), which will increase N, will result in more scattering. A diagram depicting the effect of Rayleigh scattering is shown in **Figure 2c**. What the phenomenon can look like in nature is shown in **Figure 2b**.

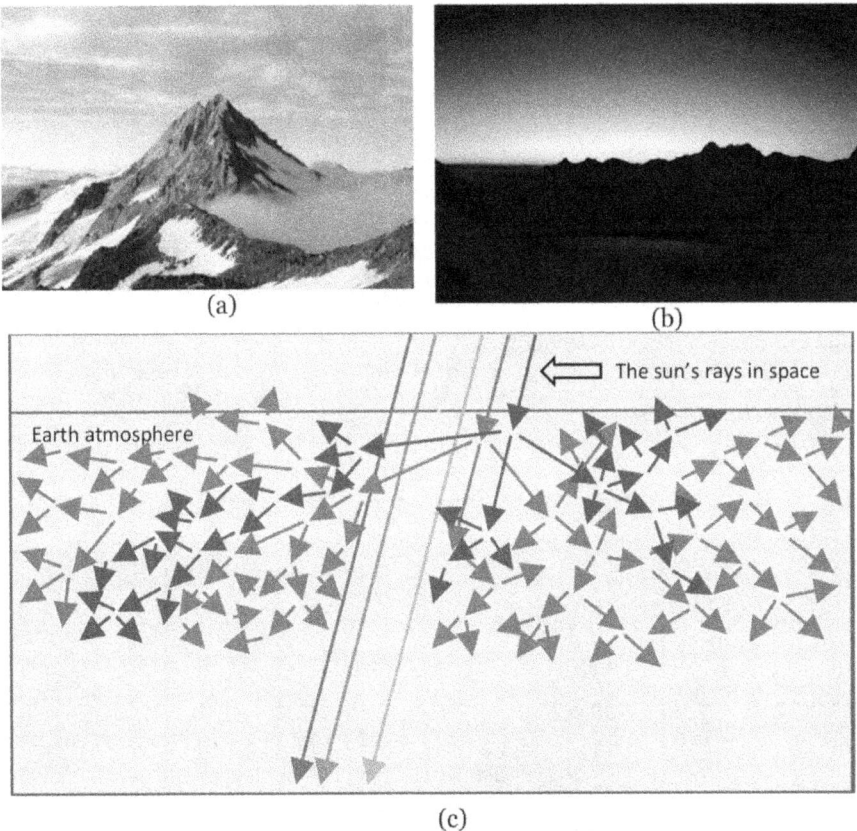

(a)

(b)

(c)

Figure 2.
Measuring the components of solar irradiance: (a) natural appearance of Mie scattering with white clouds as the Mie scattering scatters light of different wavelengths equally [8, 9]. (b) Natural appearance of Rayleigh scattering. (c) Schematic of Rayleigh scattering with colors with long wavelengths (red, yellow, and green) scattering little and the short wavelengths (blue and UV) scattering more [10].

1.3 Zenith angle effect

However, the largest influence on the irradiance that is received at a horizontal surface is the angle at which it hits the earth surface. The irradiance on the surface is proportional to $\cos(\theta z)$ where θz is the solar zenith angle [7], which is defined as the angle between the zenith and the sun. A diagram depicting this effect [11] is shown in **Figure 3**.

The solar zenith angle is a.o. dependent on the location, time of day and season, and can be calculated using SOLPOS [13].

1.4 Solar spectrum

The solar spectrum at the top of the atmosphere is very similar to the Planck's curve (black body radiation), as shown in **Figure 4**. The atmosphere influences the irradiance, and absorptions occur due to gaseous molecules such as O_2, H_2O, O_3, and CO_2, which are also depicted in the same figure.

The regions of the spectra can be subdivided in: UV: 200–400 nm, visible: 400–780 nm, and near infrared: 780–3000 nm [14].

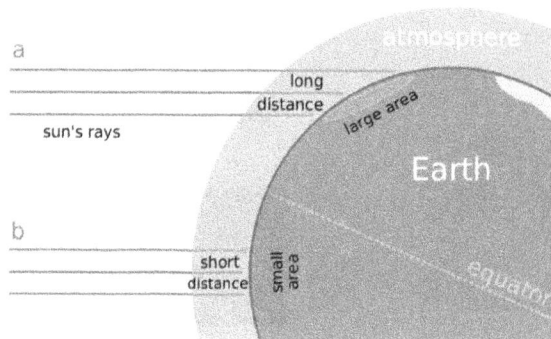

Figure 3.
Diagram [12] depicting the solar zenith angle effect, which has the largest effect on the amount of solar irradiance that is received on a horizontal surface on the earth. Shown here is a case on the equator (b) and a case close to the North Pole (a).

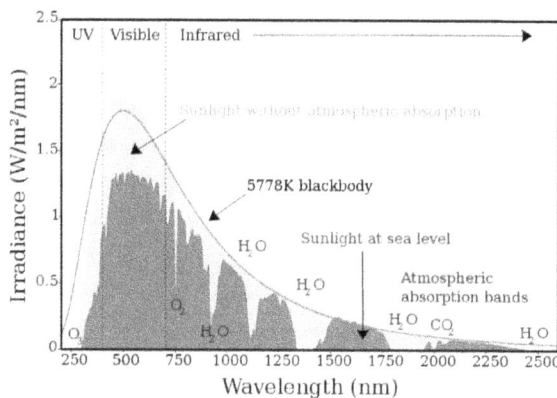

Figure 4.
The spectrum of solar radiation on earth (with slight adjustment) from: "Spectrum of Solar Radiation," by Tuvalkin is licensed under CC BY-SA 3.0.

With changing air mass (in other words, the solar zenith angle), the amount of scattering and absorption changes, and therefore, the spectrum does as well. An example of this is shown in **Figure 5** where solar spectrum curves have been calculated using the Dr. Christian Gueymard's SMARTS software [15–17]. This software allows for setting various clear sky atmospheric parameters for calculating clear sky solar spectra.

Additionally, also the cloudiness influences the spectrum that is visible on the earth surface. Three examples, for clear, cloudy, and hazy sky conditions are shown in **Figure 6**. Generally, there can be said that increase of cloudiness, just as increase of air mass (or solar zenith angle), generates a shift (in the relative spectrum) toward the red part. There will be relatively less blue and more (infra) red for more cloudy conditions than for clear sky.

1.5 Components of solar radiation

The irradiance from the sun interacts with the atmosphere, and when it arrives on the earth, it can be detected as different components (shown in **Figure 7**). These components are: direct normal irradiance (DNI), solar irradiance from the direction of the sun on a surface perpendicular to the solar rays. Diffuse horizontal irradiance (DHI) is scattered solar irradiance from the sky (except the sun) measured horizontally. Global horizontal irradiance (GHI) is the solar irradiance from the

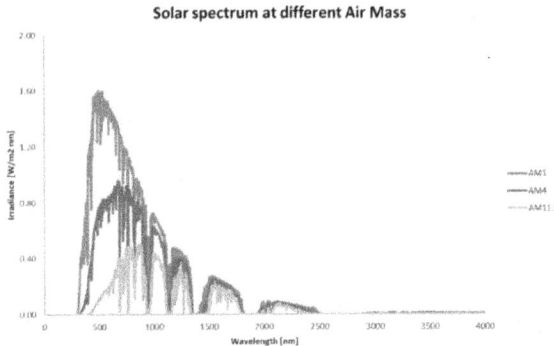

Figure 5.
The spectrum of solar radiation on earth for different air mass calculated using smarts software [15].

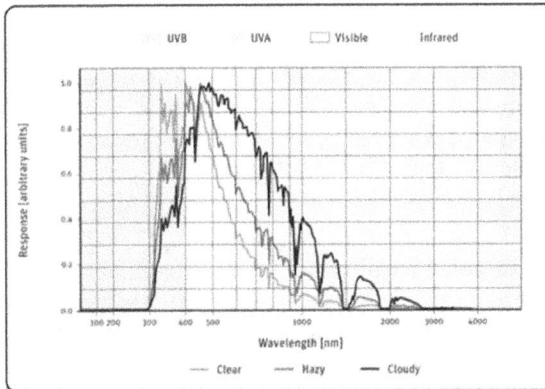

Figure 6.
The spectrum of solar radiation on earth for different sky conditions [14].

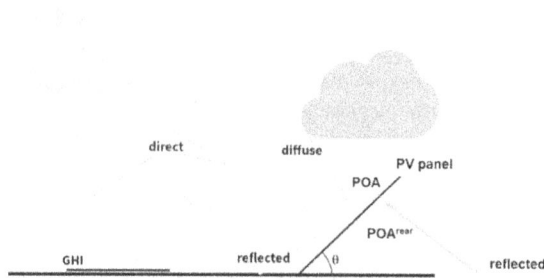

Figure 7.
The components of solar radiation: direct, diffuse, and GHI (global); and the components relevant for PV modules: POA and POA^{rear} [18].

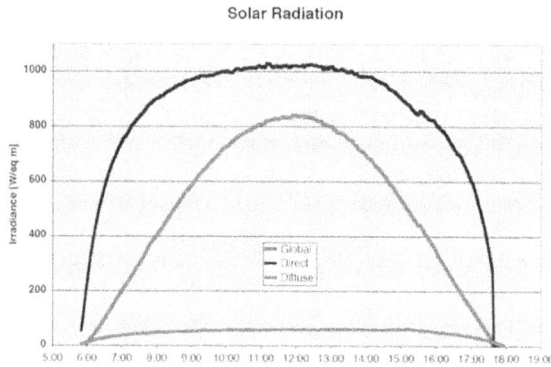

Figure 8.
On a clear day, the three components, global, direct, and diffuse, can look like this.

hemisphere above on a horizontal surface, and plane of array (POA) or global tilted irradiance (GTI) is solar irradiance incident on a tilted plane (PV panel) including radiation reflected from ground and shadowing. POArear is the solar irradiance incident on the back of the tilted plane, which is relevant for bifacial modules that can generate power from rear side irradiance.

For concentrated solar power (CSP) [19], generation of DNI is of most interest and for PV panels POA, POArear, and GHI are of interest.

The three solar components as measured on a clear day are as shown in **Figure 8**. The direct irradiance shows a typical parabola, and the diffuse is more or less constant sufficiently after sunrise or before sunset. The global irradiance is less than the direct component and is less peaked due to the solar zenith effect.

These different components can be measured with pyranometers, pyrheliometers, and solar trackers as shown in the next paragraph.

2. Measurement of solar irradiance

Solar irradiance is measured with many different radiometers depending on the desired measurement. For the UV region, radiometers are available that measure the UV-B, UV-A, total UV, or UV erythema (irradiance that causes sunburn). There are thermopile radiometers, also called pyranometers, that measure from 280 to roughly 2800 nm. Also, there are photodiode variants of the pyranometer that typically measure from 300 to 1000 nm. And there are pyrgeometers that measure in the infrared.

For PV applications, the most relevant radiometers are the thermopile and silicon pyranometers as well as the pyrheliometers.

A pyranometer, its name derived from Greek (Πυρά – ἄνω – μετρ-έω, pyra – ano – metreo, fire – heaven – measure), is a thermopile-based instrument, with broadband black coating, which measures the net radiation coming from a 180° half dome above the instrument, allowing measurement of GHI, POA, or DHI. The pyrheliometer, its name derived also from Greek (Πυρά – ἥλιος – μετρ-έω, pyra – helios – metreo, fire – sun – measure) consists of a collimation tube with an opening angle of 5° and slope angle of 1°, which blocks the light not coming from the direction of the sun, allowing a measurement of the solar beam (or DNI) when pointed at the sun.

Examples of these instruments measuring the different solar components are given in **Figure 9**, where it is shown that pyranometers can be used to measure GHI, POA, and DHI. For measuring DHI, both a shadow ring and a solar tracker with shading ball can be used. The shadow ring is a smaller investment but would have to be adjusted manually depending on the change of the sun elevation, which changes over the year. The solar tracker tracks the sun automatically without manual adjustment. For measuring the DNI also a solar tracker is used to point the pyrheliometer at the sun.

Solar irradiance measurement is important in many fields such as meteorology, climatology, building automation, and material research. However, the fastest growing application is in solar energy.

Solar energy applications are both in concentrated solar and in photovoltaic energy generation. For concentrated solar, the sunlight is concentrated to heat a small area, which generates electricity as a conventional power plant. The

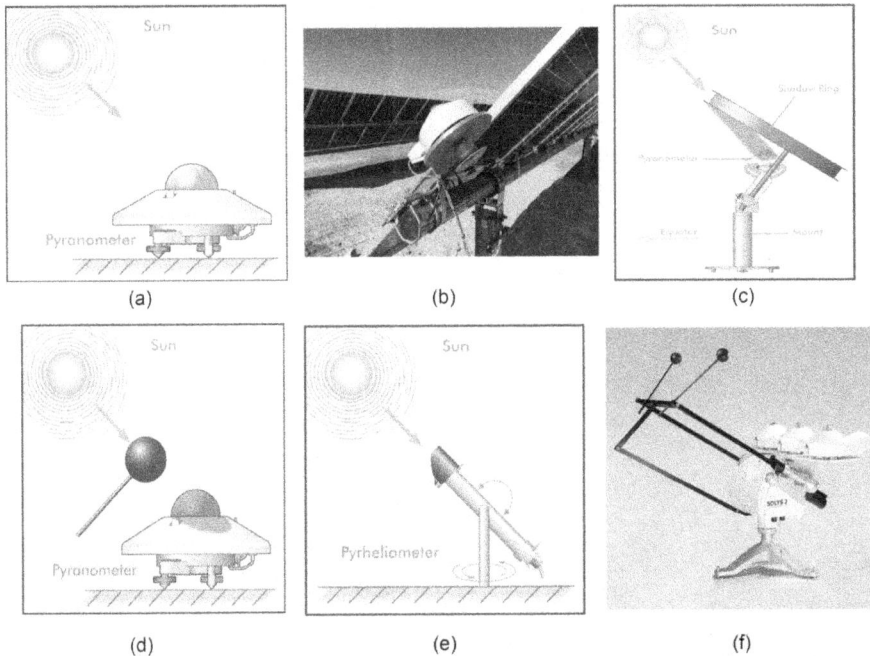

Figure 9.
Measuring the components of solar irradiance: (a) GHI with pyranometer, (b) POA and POArear with a tilted pyranometer next to bifacial modules, (c) DHI with shadow ring and pyranometer, (d) DHI with shading ball on tracker, (e) DNI with pyrheliometer on a tracker, (f) SOLYS gear drive sun tracker with shading balls, pyrheliometer, and ventilated pyranometers [14].

measurements of importance for concentrated solar are the three solar components with an emphasis on the DNI.

For photovoltaic energy generation, the sunlight is used for direct conversion to electricity in the modules. The measurements of importance for photovoltaics are POA and POA_{rear} for the calculation of performance ratio. Additionally, other components can also be importance, such as GHI for comparison of data to local meteorological stations or satellite observations and also albedo measurements for bifacial plants.

2.1 Solar monitoring using IEC 61724-1

From the IEC 61724-1 [18], there are a number of classes of solar monitoring: class A and B (going from high to low accuracy). Requirements regarding the sampling and recording intervals for irradiance measurements are as follows:

	Class A	Class B
Max. sampling interval	5 s	1 min
Max. recording interval	5 min (1 min recommended)	15 min

The data have to be stored with a date and timestamp in either local or universal time.

The number of POA or GHI sensors recommended depends on the size of het power plant as well as the desired class of the monitoring system. For example, 5–40 MW would require two sensors and 500–700 MW six sensors.

Depending on the desired accuracy class, there are three possible choices of sensors: irradiance sensors such as thermopile or photodiode pyranometers, matched irradiance sensors such as reference cells.

Before using the measurement data in performance ratio calculations, a data quality check needs to be done where invalid data are identified and filtered out. Additionally, missing data will have to be treated in a certain way such as: taking the data from before and after the missing data, using a predefined method from a contract, or leaving the interval out of the analysis.

2.2 Application in photovoltaics

In photovoltaics, there are many applications for measuring solar irradiance with one of the first fields where measurements are important in the technology research (see **Figure 10**) [20].

(a)   (b)

Figure 10.
(a) GHI and (b) POA technology research using pyranometers and pyrheliometers.

Improvements in mass-produced PV technologies, such as crystalline silicon cells, are incremental; each step is small, but the total gain can be large. For example, two different solutions may show efficiencies of 20% and 22% (10% improvement) under controlled laboratory test conditions. However, the irradiance needs to be measured with an uncertainty of better than 2% to be sure that the efficiency improvement is genuine. Laboratory testing under indoor conditions is not enough. The performance of cells and modules needs to be verified in the real-world settings under varying meteorological and sky conditions compared with "reference" quality solar radiation measurements. Therefore, research institutes are equipped with scientific-level pyranometers, and in many cases, a complete solar monitoring station.

Apart from research, irradiance measurements are also used for quality control if a manufacturer or a supplier wants to know if the performance of their PV cells or modules does not vary by more than (for example) 5%, they need to test samples from production batches and measure the solar radiation with a significantly lower uncertainty. Verification of published specifications of equipment the manufacturer or an independent test laboratory, needs reference quality measurements.

2.3 Performance ratios

There are many stages in the development of a utility-scale solar power plant, and throughout its operational life and in all of them, the performance ratio (PR) is of key importance. In the early stages, this is an estimation or prediction but, in the later stages, it uses real plant monitoring data. PR is the ratio between the final (actual) yield of a solar power generating system and its reference (design) yield over a defined period of time. For PV systems, the actual yield of the generated AC power is easily measured accurately at the grid connection. The reference yield is the expected power produced by irradiance on the PV modules; the solar energy received by the panels multiplied by the efficiency of the conversion to electrical energy and which should include the inverter efficiency, cabling, and connection losses. Performance ratios (if defined and monitored in the same way) can be used to compare solar plants at any locations: a well-designed, -installed, and – maintained solar park in the northern latitudes could have a better PR than an averagely designed and installed solar park near the Equator (although the latter receives far more energy from the sun). How to calculate the performance ratio is defined in IEC 61724-1 [18] (of which a new version has come out in the summer of 2021):

$$PR = \left(\sum_k P_{out,k} \times \tau_k \right) / \left(\sum_k \frac{P_0 \times G_{i,k} \times \tau_k}{G_{i,ref}} \right) \tag{5}$$

Where P_{out} is the power out, t_k is the time period, P_o is the power out under reference conditions, $G_{i,k}$ is the in-plane irradiance (POA) during time period k, and $G_{i,ref}$ is the POA during reference conditions. For reduction of variation in this PR for the different seasons, a temperature-corrected formula is also provided in IEC-61724-1.

2.4 Site prospecting

A solar energy project starts with site prospecting—finding the optimal location for the plant. Solar energy resource maps are widely available and are often used to derive the potential for solar electricity in a particular region. These are usually generated from models using up to 10 years of satellite data and ground-based meteorological observations (often widely spaced and not very accurate) and inter-polation. However, the map data are not good enough quality, and the spatial scale too large, to provide a reliable basis on which to make technology and investment decisions for a power plant. Due to local climate and topographical differences, relatively small changes in location can result in a gain, or loss, of hundreds of annual sunshine hours per year, particularly in mountainous and coastal areas and for islands. Other meteorological factors also have to be taken into account such as cloud, fog, and rain that reduce the amount of energy produced. Local infrastruc-ture issues also play an important part, site access for construction and mainte-nance, and proximity to a grid to feed in the generated power.

The information above would allow potential sites to be short-listed. However, in order to decide on which are the most economically attractive, and to select the optimum power generation technology for a site, high-quality ground-based irradi-ance measurements over at least a year are required. Meteorological measurements by an automatic weather station are also needed and allow comparison with histor-ical data to ascertain if it is a typical year. The parameters are usually wind speed and direction, precipitation, air temperature, and relative humidity, GHI measure-ments by pyranometers can be used to validate and "train" for that specific location the GHI estimates derived from satellite data models. POA irradiance cannot be accurately derived with a model at a suitable level of uncertainty in order to make investment decisions, this needs local tilted pyranometers.

Experienced investors want the lowest uncertainty of the on-site solar resource data, generating equipment performance and proven reliability, before making decisions on the locations for solar energy plants and on the most effective PV system types to use. Errors in the solar radiation measurements can significantly impact upon the difference between predicted and achieved return on investment.

The estimated performance ratio indicates the potential profitability of a PV plant, and high-quality, reliable local solar radiation data are critical to the bank-ability of projects. See **Figure 11** for an example of prospecting.

2.5 Plant design

Good solar plant design optimizes yield and reduces losses, resulting in a high PR [21]. The design and the equipment selected are heavily influenced by the environ-ment surrounding a solar energy plant in terms of irradiation, sun elevation paths, shading (by mountains, trees, buildings, clouds), temperature ranges, precipitation, wind, pollutants, and soiling.

These environmental factors are important information retrieved during the site prospecting phase: most are naturally occurring and cannot be easily changed, and

Figure 11.
Solar prospecting of GHI and tilted GHI or POA.

they influence the mechanical and electrical design but also the expected maintenance required during plant operation. Often, stakeholders have a preferred list of
suppliers; companies and products with good quality, performance, reliability, and
cost over the lifetime. Higher quality and performance instruments will in general
provide a more reliable long-term performance ratio, with lower uncertainty. By
using accurately measured solar irradiance and the back panel temperature-
corrected performance ratio, two critical environmental parameters for PV systems
are taken into account, both for the reference and final yields. A mean annual
temperature 2°C higher than the value used in the reference calculations can drop
the PR by 1%. Accurate local measurements also enable PR to be used over shorter
time periods, for instance, monthly.

2.6 Installation and commissioning

Following the design scheme as closely as possible during construction and
installation is key to reaching the projected reference PR. An initial period of
operation, from several weeks to months, is used to calculate the commissioning PR.
From this a target PR is derived. The contractually agreed target performance ratio
(sometimes called the guaranteed performance ratio) is often slightly lower than
the final PR, to allow O&M parties to correct faults and restore interrupted operation. A checklist for this is provided by the 2015 International Finance Corporation
Project Developer's Guide, Utility-Scale Solar Photovoltaic Power Plants [22]. By
showing a high performance ratio after the initial building, commissioning, and
operation time periods, EPCs can show their ability make well-performing PV
plants. Plants like these will generate a higher selling price on the secondary market
and reduce the future risk for the buyers. But, to do this requires suitably quality
irradiance data.

2.7 Plant operation

The monitoring of a solar power plant [20, 21] is a complex process with many
stages, from solar energy input to grid electrical power output. For all these stages,
separate instruments and associated software are available to monitor the process.
During the first few years, the final performance ratio of the plant is determined,
operating efficiency is maximized, and the true O&M costs can be assessed, leading
to an overview of the financial return on the investment. Of course, this includes
the quality of the solar irradiance data. By maintaining yield and availability at high
levels at modest costs, O&M parties can show their added value in optimized
operating and maintenance policies. A high performance ratio shows the quality of
their work. Gradual changes in efficiency compared with the local irradiance measurements may indicate dirty panels, so cleaning actions can be scheduled. More
sudden changes may indicate a defective section, a cable and connection problem,
or an inverter issue, so further service actions are required to find the problem. See
Figure 12 for examples of operational monitoring.

2.7.1 Output forecasting

Using high-quality solar radiation monitoring at the plant, a dataset of performance can be built up, allowing more accurate forecasting of the future energy yield
and financial returns [22, 23]. Real-time measurements and a historical database can
be used in conjunction with satellite data and weather forecasts as inputs to now-
casting models for the output of the plant in the coming hours. This is of particular

(a)   (b)

Figure 12.
Operational monitoring of (a) concentrated photovoltaics (CPV) where a pyrheliometer is used to measure the DNI (inset) and (b) PV in Italy with POA and GHI monitoring.

interest to grid operators, as other power generation sources cannot be switched instantly when clouds pass over the solar energy plant.

2.7.2 Reinvestment

Over time, the plant PR will degrade and a business case for refurbishing can be made involving investment in new equipment: replacement panels, inverters, transformers, cabling, etc. Studies such as the Compendium of Photovoltaic Degradation Rates from NREL [24] show that the performance of PV panels commonly reduces by 0.5–1% per annum. A refurbishment might take place after 20 years of operation, or when a power plant is sold on, and the replacement will normally have better specifications and performance than the original equipment.

PV panels have a wide field of view and must be positioned in such a way as to receive the maximum amount of solar radiation at the desired time of year. Depending on the local conditions, as well as the land price per area, a cost/benefit decision can be made whereby they are usually installed at a fixed angle. In this case, a pyranometer is required, tilted in the plane of array of the fixed panels to measure the solar radiation received by the modules. As this pyranometer is part of the plant performance monitoring system, a second POA unit is often fitted for redundancy and back-up in case of recalibration of one of the units. If the panels vary significantly in their azimuth and/or zenith angles, for example, where the plant is across a valley or an undulating hillside, additional pyranometers are needed. As the plant size increases, it takes time for clouds to move across, and some parts will be in shadow, and others will be in sunshine, requiring more irradiation measurement points. To maximize the use of the available solar energy, PV panels are often installed on mountings that move to follow the sun during the day, either by rocking about a single axis or on a two-axis sun tracker. POA pyranometers can then be mounted to the module frames (**Figure 12**).

3. Pyranometer measurement principles

Both instruments consist of a thermal element (either a thermopile or Peltier) built into a metal body [25, 26]. A thermopile consists of multiple thermocouples, consisting of two different metals connected at a "hot" and "cold" junction that generate a voltage when subjected to a temperature difference. Solar irradiance heats the black coating of the instrument, which also results in a temperature increase of the "hot" junctions of the thermopile. The "cold" junctions of the thermopile are in contact with the colder metal body. The resulting temperature difference between the hot and cold junctions generates a voltage through the

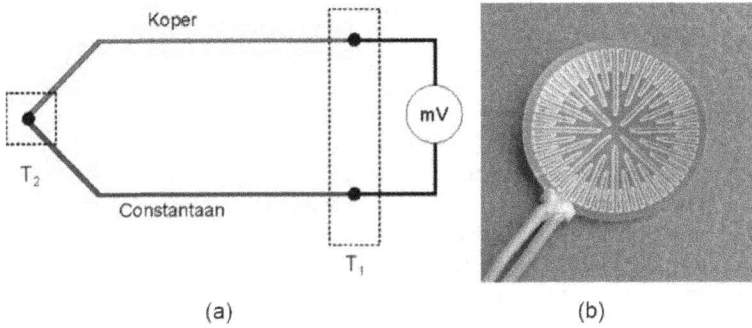

Figure 13.
(a) Measurement principle of a thermocouple measuring a temperature difference and (b) layout of multiple thermocouples as used in a pyranometer.

Seebeck effect. This effect is linear with the temperature difference and the magnitude determined by the Seebeck constant of the materials of the thermopile. With either an external datalogger or an onboard ADC for digital sensors (e.g., the Kipp & Zonen SMP pyranometers), the voltage is measured, and the irradiance can be obtained.

Figure 13 shows the basis of a thermopile pyranometer with a thermocouple diagram (a) and the combination of multiple of these thermocouples to form a thermopile as shown in photograph (b). The physical effect that generates a voltage across the thermocouple (and thermopile) is the Seebeck effect:

$$V_{Seebeck} = \alpha \cdot \Delta T \tag{6}$$

Where $V_{Seebeck}$ is the Seebeck voltage, α the Seebeck coefficient, and ΔT the temperature difference. How the thermopile functions within a pyranometer to measure solar irradiance is shown in **Figure 14**.

Where the sunlight is transmitted by the dome onto the black coating on top of the thermopile (or in some cases, a Peltier showing the same voltage effect). The black coating heats up as does the underlying thermopile hot side. The thermopile cold side is connected to the heat sink and remains at ambient temperature, and this generates a temperature gradient over the thermopile. Because of this temperature gradient, the Seebeck effect generates a voltage difference at the thermopile output.

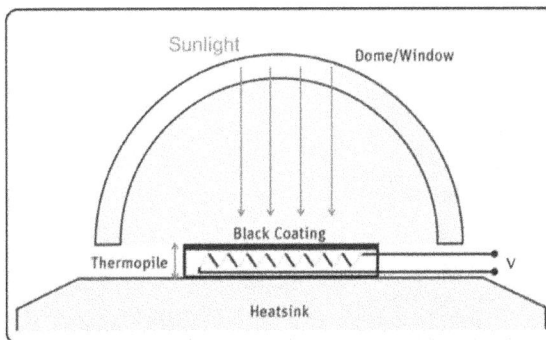

Figure 14.
Depiction of a pyranometer with the dome, black coating, thermopile, and heat sink, with sunlight incident on the pyranometer.

The measurement of the voltage allows for a measurement of the irradiance onto the pyranometer.

3.1 Pyranometer classification

Pyranometers are classified according to ISO 9060:2018 (in order of increasing quality: "class C," "class B," and "Class A"). The higher the instrument class, the better the time response, zero offsets, nonstability, nonlinearity, directional response, clear sky global horizontal irradiance spectral error, temperature response, tilt response, and additional signal processing errors. Two other additional

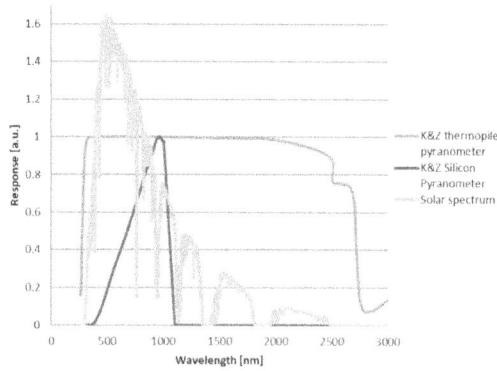

(a)

(b)

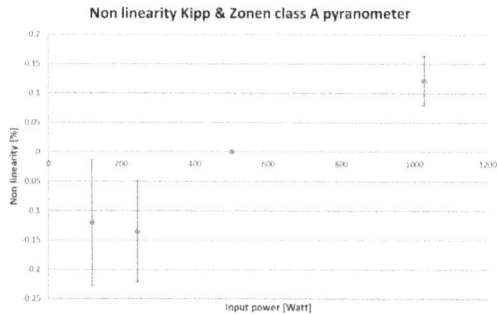

(c)

Figure 15.
(a) Kipp & Zonen thermopile pyranometer and silicon pyranometer spectral response plotted with the solar spectrum; (b) recalibration data of a CM22 Kipp & Zonen pyranometer located at NREL over 20 years [27]; and (c) nonlinearity of a class A pyranometer.

classifications are spectrally flat, for a constant spectral response from 350 nm to 1500 nm, and fast response, for a response time of less than 0.5 seconds.

The stability of a Kipp & Zonen CM22 pyranometer is shown in **Figure 15b** and shows a very stable instrument over nearly 20 years at NREL in Golden USA, based on publicly available data of recalibrations [27]. The nonlinearity of a class A Kipp & Zonen pyranometer is shown in **Figure 15c** and shows a nonlinearity very of approximately 0.13% at 1000 W/m^2.

Both thermopile pyranometers and silicon pyranometers can be classified using the ISO 9060 standard. The most significant difference between the two is the spectral response as shown in **Figure 15a**, with the thermopile pyranometer having a spectral response, combined from measured black coating absorptance, glass transmission, and Fresnel reflections,) that is very flat from 0.3 to 2.7 µm, encompassing 99.8% of the solar spectrum. The silicon pyranometer has a varying response over a smaller range of the solar spectrum, which results in a somewhat higher spectral error for clear sky global horizontal irradiance.

3.2 Pyranometer calibration and uncertainty

The pyranometers and pyrheliometers are calibrated to the World Radiometric Reference (WRR). This reference is recognized by SI as a conventional standard for solar radiation measurements and is maintained by the World Standard Group (WSG) consisting of absolute cavity pyrheliometers (**Figure 16**). At Kipp & Zonen, we use an absolute cavity pyrheliometer (PMO-6) and a pyrheliometer (CHP1) calibrated against the WSG at the International Pyrheliometer Comparisons (IPC) every 5 years and checked at the National Pyrheliometer Comparison (NPC) at the National Renewable Energy Laboratory (NREL) in Golden Colorado annually.

With these pyrheliometers, we calibrate the reference pyranometers according to the Alternating Sun Shade (ASS) in ISO 9846. The ASS method as compared with the Continuous Sun Shade (CSS) method has the advantage of a lower calibration uncertainty, as well as a lower influence during calibration by offsets varying with atmospheric conditions [29]. The setup for ASS, which alternatingly exposes and shades pyranometers and simultaneously measures the DNI with an absolute cavity pyrheliometer, is shown in **Figure 17a**. To transfer the calibration of the reference to the production instruments, the ISO 9847 is followed using the direct beam Kipp & Zonen method with a lamp as a light source (see **Figure 17b**).

Figure 16.
World standard group (WSG) of absolute cavity pyrheliometers at WRC PMOD in Davos Switzerland [28].

(a)   (b)

Figure 17.
(a) rotatable shading disc setup for alternating sun shade (ASS) calibration setup and (b) diagram of indoor calibration of test pyranometer next to reference pyranometer according to ISO 9847.

The uncertainty of pyranometers in the field is a combination of the calibration uncertainty and the interaction of the pyranometer parameters with the site conditions.

The calibration uncertainty is stated on a calibration certificate and consists usually of a combination of the uncertainty of the reference pyranometer combined with the uncertainty of transferring the calibration from the reference to the pyranometer to be calibrated. Calibration uncertainties (with a coverage factor of k = 2, 95% confidence interval) of pyranometers can be as low as below 1% depending on the pyranometer parameters as well as the calibration process.

The uncertainty of a pyranometer in the field is dependent not only on the calibration uncertainty and instrument quality, but also on the site conditions such as location, time, sky conditions, and instrument maintenance. This can be calculated with the help of the Suncertainty app for different Kipp & Zonen pyranometers as a function of site conditions [30].

4. Alternative measurement methods

Alternative methods for measuring irradiance are reference cells or satellites. Reference cells, if completely identical to the PV panel, allow one to determine the energy the PV panel collects. However, to measure the broadband irradiance, their accuracy is less good than that of pyranometers. Furthermore, some research questions the stability of non-monocrystalline silicon reference cells [31]. Satellite-derived irradiance data have generally a lower accuracy, especially for short time frames, but can be complementary to ground measurement. Satellites perform better under clear sky conditions than cloudy conditions [32], and for longer time-scales, the uncertainty is reduced [33].

5. Conclusions

Solar irradiance is of utmost importance for PV energy generation and can be affected in different ways. To a lesser extent, it is the variation of sunlight reaching the top of the atmosphere due to the sun cycle as well as the variation in sun-earth distance. To a larger extent, the atmosphere creates a variation by scattering at particles or clouds. Additionally, the solar zenith angle, varying with season and solar time, is of large influence.

POA solar irradiance can be measured with pyranometers, silicon pyranometers, and reference cells. These measurements are necessary for PV site prospecting, design, and operation.

For solar energy applications, pyranometers have the lowest uncertainty for GHI and POA broadband irradiance measurements, and aside from absolute cavities, pyrheliometers are the most accurate way to measure DNI.

Acknowledgements

The author acknowledges the help of Clive Lee, Kipp & Zonen, Dr. Joop Mes, Kipp & Zonen, and Dr. Ing. Mohammadreza, NTNU Alesund.

Conflict of interest

The author declares a conflict of interest as he is working as a scientist at Kipp & Zonen, pyranometer manufacturer.

Author details

Marc A.N. Korevaar
Kipp and Zonen, Delft, Netherlands

*Address all correspondence to: marc.korevaar@otthydromet.com

IntechOpen

References

[1] Pinker RT. Surface radiative fluxes. In: Njoku EG, editor. Encyclopedia of Remote Sensing. New York: Springer; 2014. DOI: 10.1007/978-0-387-36699-9_199

[2] Kopp G, Lean JL. A new, lower value of total solar irradiance: Evidence and climate significance. Geophysical Research Letters. 2011;**38**:L01706. DOI: 10.1029/2010GL045777

[3] arXiv:1601.05397v1 [astro-ph.SR]. https://doi.org/10.48550/arXiv.1601.05397

[4] Solar constant data. 2021. Available from: https://spot.colorado.edu/~koppg/TSI/ [Accessed: February 8, 2021]

[5] Vignola F, Michalsky J, Stoffel T. Solar and Infrared Radiation Measurements. 1st ed. Boca Raton: CRC Press; 2012. DOI: 10.1201/b12367

[6] Part 2: Solar Energy Reaching The Earth's Surface – ITACA. Available from: www.itacanet.org. http://www.itacanet.org/the-sun-as-a-source-of-energy/part-2-solar-energy-reaching-the-earths-surface/#2.1.-The-Solar-Constant [Accessed: February 21, 2022]

[7] Solar zenith angle. Available from: https://en.wikipedia.org/wiki/Solar_zenith_angle [Accessed: May 5, 2022]

[8] Hahn, David W. Light Scattering Theory (PDF), 2009, University of Florida. Available from: http://plaza.ufl.edu/dwhahn/Rayleigh%20and%20Mie%20Light%20Scattering.pdf [Accessed: August 13, 2021]

[9] Mie scattering (Wikipedia). 2021. Available from: https://en.wikipedia.org/wiki/Mie_scattering [Accessed: August 13, 2021]

[10] Rayleigh scattering (Wikipedia). 2021. Available from: https://en.wikipedia.org/wiki/Rayleigh_scattering [Accessed: August 13, 2021]

[11] The cosine effect. Available from: https://www.itacanet.org/the-sun-as-a-source-of-energy/part-2-solar-energy-reaching-the-earths-surface/#2.2.-The-Cosine-Effect [Accessed May 5, 2022]

[12] Halasz P. Effect of the Earth's shape and atmosphere on incoming solar radiation. 2007. Available from: http://commons.wikimedia.org/wiki/File: Oblique_rays_03_Pengo.svg, licensed under CC BY-SA 3.0 [Accessed: August 2, 2021]

[13] SOLPOS calculator. Available from: https://midcdmz.nrel.gov/solpos/solpos.html [Accessed: February 21, 2022]

[14] Rosemann R. A Guide to Solar Radiation Measurement, from Sensor to Application. Kipp & Zonen; 2011

[15] SMARTS: Simple Model of the Atmospheric Radiative Transfer of Sunshine. 2006. Available from: https://www.nrel.gov/grid/solar-resource/smarts.html [Accessed: August 2, 2021]

[16] Gueymard C. Parameterized Transmittance Model for Direct Beam and Circumsolar Spectral Irradiance. Solar Energy. 2001;**71**(5):325-346

[17] Gueymard C. SMARTS, A Simple Model of the Atmospheric Radiative Transfer of Sunshine: Algorithms and Performance Assessment. Professional Paper FSEC-PF-270-95. Cocoa, FL: Florida Solar Energy Center; 1995

[18] IEC. IEC 61724-1 Photovoltaic System Performance – Part 1: Monitoring. 2021

[19] Concentrated solar power [Internet]. Available from: https://en.wikipedia.org/wiki/Concentrated_solar_power [Accessed: May 5, 2022]

[20] Rummel S et al. PV cell and module performance measurement capabilities at NREL. In: National Center for Photovoltaics Program Review Meeting. Denver, Colorado; 1998

[21] Gueymard CA, Wilcox SM. Assessment of spatial and temporal variability in the US solar resource from radiometric measurements and predictions from models using ground based or satellite data. Solar Energy. 2011;**85**:1068

[22] International Finance Corporation. Utility-Scale Solar Photovoltaic Power Plants: A Project Developer's Guide. 2015. Available from: https://www.ifc. org/wps/wcm/connect/topics_ext_conte nt/ifc_external_corporate_site/sustainab ility-at-ifc/publications/publications_ utility-scale+solar+photovoltaic+power +plants [Accessed: August 3, 2021]

[23] van Sark W, Reich BMNH. World Renewable Energy Forum (WREF) Congress XII. Denver, CO; 2012. pp. 1-6

[24] Nordmann T et al., Tech. Rep. (Photovoltaic Power Systems Programme (PVPS), IEA. 2014

[25] Moll WJH. A thermopile for measuring radiation. Proceedings of the Physical Society of London. 1922;**35**:257

[26] Bener P. Untersuchung über die Wirkungsweise des Solarigraphen Moll-Gorczynski. Archiv für Meteorologie, Geophysik und Bioklimatologie, Serie B. 1950;**2**:188

[27] MIDC/AIM Calibration Database [Internet]. 2021. Available from: https:// midcdmz.nrel.gov/apps/fmpxml.pl? type=calib;find=010034 [Accessed: August 2, 2021]

[28] World Radiation Center. 2021. Available from: https://www.pmodwrc. ch/ [Accessed: August 2, 2021]

[29] Philipona R. Underestimation of solar and diffuse radiation measured at Earth's surface. Journal of Geophysical Research. 2002;**107**:4654

[30] Kipp & Zonen B.V. Suncertainty [Mobile Application Software]. 2016. Available from: https://play.google.com/ store/apps/details?id=com.kippzonen. suncertainty&hl=nl

[31] Woyte A et al. Monitoring of photovoltaic systems: Good practices and systematic analysis. In: PV Solar Energy Conference. 2013

[32] Greuell W, Meirink JF, Wang P. Retrieval and validation of global, direct, and diffuse irradiance derived from SEVIRI satellite observations. Journal of Geophysical Research: Atmospheres. 2013;**118**:2340-2361. DOI: 10.1002/jgrd.50194

[33] Suri M, Cebecauer T. Uncertainty of satellite-based solar resource data. In: SolarPACES Conference. Bejing, China; 2014 Available from: solargis.com

Chapter 3

Modelling of Solar Radiation for Photovoltaic Applications

David Afungchui, Joseph Ebobenow, Ali Helali
and Nkongho Ayuketang Arreyndip

Abstract

This chapter explores the different ways in which solar radiation (SR) can be quantified for use in photovoltaic applications. Some solar radiation models that incorporate different combinations of parameters are presented. The parameters mostly used include the clearness index (K_t), the sunshine fraction (SF), cloud cover (CC) and air mass (m). Some of the models are linear while others are nonlinear. These models will be developed for the estimation of the direct (H_b) and diffuse (H_d) components of global solar radiation (H) on both the horizontal and tilted surfaces. Models to determine the optimal tilt and azimuthal angles for solar photovoltaic (PV) collectors in terms of geographical parameters are equally presented. The applicable, statistical evaluation models that ascertain the validity of the SR mathematical models are also highlighted.

Keywords: Global, Direct, Diffuse, Solar Radiation, Modelling, Linear models, Nonlinear models, Least square method, statistical evaluation models

1. Introduction

Solar radiation is essentially a flux of photons originating from the sun and radiating in all directions of space. These photons exhibit electromagnetic wave properties and travel at the speed of light over an average distance of about 149.4 million km to reach the earth's surface while suffering diverse attenuations from the components of space and the earth's atmosphere.

Many devices are being employed to measure SR but the scope of such measurements over space and time is limited. As a consequence, it is mandatory to develop alternative heuristic models to qualify and quantify solar radiation.

Data on global solar radiation (GSR) is readily available in most meteorological stations around the world but data on the diffuse and beam components of SR is rare and needs to be estimated by alternative means. Measurements of SR are mostly done on horizontal surfaces while real-time solar PV receivers require tilting from the horizontal position for optimal harvesting of the SR [1]. Information on both the direct and diffuse components of SR is necessary to accurately characterise the irradiance intercepting a solar collector or receiver.

GSR is short wavelength radiation that can characteristically be either broadband or spectral. From this premise, SR is modelled using either broadband or spectral models. Besides, satellite-based models have also been developed. The broadband models are suitable for ground-based measurements. A plethora of submodels, with varying levels of complexity, now exist and will be presented in the subsections that follow.

The general trend over the past decades pioneering with the work of John K. Page [2], is the development of models which have been severally tested and improved upon. The common approach in the models is to predict either the diffuse and/or the direct SR components from measured GSR data. Alternatively, some models use meteorological parameters like temperature, sunshine hours and relative humidity, together with the GSR data to predict the direct or diffuse components.

Except in the subsection(s) where we treat the aspect of tilt angle, every occurrence of radiation henceforth will be considered to mean radiation measured (or predicted) regarding a horizontal surface.

To ascertain the accuracy of the models, some statistical tools for the evaluation of the models have been presented. These include the mean bias error (MBE), the root mean square error (RMSE) and t-statistics [3–5]. A. S. Angstrom [3] disclosed that these statistical tools collectively combine to establish the consistency of the models.

This chapter will be organised as follows: After this introduction, we will present in the next section the statistical tools applicable for testing of the model's performance. This is followed in Section 3 by an exploration of the different approaches used in modelling solar radiation. Given that our emphasis is on photovoltaic technology, we present in the last section the modelling of tilt and azimuth angles in connection with solar photovoltaic energy applications. This is followed by the concluding remarks on the chapter.

2. Statistical evaluation methods for photovoltaic solar radiation models

The prediction efficiency of the models being presented in this chapter needs testing to ensure their validity and reliability. This is achieved using some statistical tools. These include: the mean bias error (MBE), the mean relative error (MRE), the root mean square error (RMSE) and the t-statistic (t-stat) error [5].

2.1 Mean bias error

The MBE is expressed as [6, 7]:

$$MBE = \frac{1}{k} \sum_{i=1}^{k} (y_i - x_i) \tag{1}$$

where x_i is the ith observed value, y_i the ith predicted value and k the total number of observations.

The mean bias error (MBE) is a pointer of the long-term performance of a correlation. This is achieved by calculating the real deviation between the predicted and measured values term wise. Ideally, an MBE value of zero is the best indicator. A positive MBE indicates an over-estimation while a negative MBE indicates under-estimation. Under practical conditions, vanishingly small MBE values are desirable for a good model's performance.

2.2 Root mean square error (RMSE)

The RMSE is expressed as [5–7]:

$$RMSE = \sqrt{\frac{1}{k} \sum_{i=1}^{k} (y_i - x_i)^2} \tag{2}$$

The root mean square error (RMSE) is determinant for the short-term performance of a regression model. The RMSE estimates the differences between observed and predicted results of some quantity being modelled, which in this case is the solar radiation. RMSE is a good measure of precision and its value is always positive, representing zero in the ideal case [6].

2.3 Mean relative error

The mean relative error (MRE) tests the linearity between the measured and the estimated values. It is expressed in the form [8];

$$MRE = \frac{1}{k} \sum_{i=1}^{k} \left| \frac{y_i - x_i}{x_i} \right| \tag{3}$$

The MRE is always positive, approaching zero in the ideal case.

Each statistical assessment tool considered alone might not be a sufficient pointer of a model's validity. It is likely to have a large RMSE value and at the same time a small MBE (a large scatter about the line of estimation). It is also possible to have a relatively small RMSE and a relatively large MBE (consistent over-estimation or underestimation).

Although these statistical indicators generally provide a reasonable tool for model performance, they do not objectively indicate whether the model's estimates are statistically significant. An additional statistical indicator, the t-statistic can be used.

2.4 The t-statistical method

Stone [9] demonstrated that the MBE and the RMSE separately do not represent a reliable assessment of the model's performance and can lead to the false selection of the best model from a set of candidates. To determine whether or not the equation estimates are statistically significant, Stone [9] proposed the t-stat expressed as:

$$t - stat = \sqrt{\frac{(n - 1)MBE^2}{RMSE^2 - MBE^2}} \tag{4}$$

T-stat values are always positive and vanishingly small values indicate a better model's performance. The parameter, n, represents the numbers of observations and corresponds to the twelve months (n = 12) of the year if average monthly measurements are used. This statistical indicator compares models and at the same time indicates whether the model's estimates are statistically significant at a particular confidence level [9, 10]. Consequently, the t-statistic is used in combination with the RMSE and MBE to give a more reliable prediction [11]. After the estimation of a coefficient, the t-statistic for that coefficient expresses the ratio of the coefficient to its standard error.

3. Approaches in solar radiation modelling

3.1 Introduction

Before reaching the earth's surface, SR suffers some of the attenuations from air particles, aerosols, water vapour and clouds. This causes the GSR to be split into three components: the reflected, the direct (or beam) and the diffuse SR components.

Several forms of SR data exist, which could be used for a variety of purposes in the design and development of solar PV systems. Daily data is often available and hourly radiation data can be estimated from available daily data.

The monthly average daily extraterrestrial radiation on a horizontal surface is expressed as [12, 13]:

$$H_0 = \frac{(24)(3600)}{\pi} G_0 \left(cos\varphi \, cos\delta \, sin\,\omega_s + \frac{\pi}{180}\omega_s \, sin\varphi \, sin\delta \right) \tag{5}$$

Here, G_0, is the extraterrestrial radiation (SR incident on the outside of the earth's atmosphere) given by:

$$G_0 = I_{sc} \left(1 + 0.034 \, cos \left(\frac{360 \, n_{day}}{365.25} \right) \right) \tag{6}$$

Where I_{sc} is the solar constant and has a value of $1.367 kWm^{-2}$ [14], φ is the latitude of the site, δ is the solar declination angle, ω_s is the sunshine hour angle for the month and n_{day} is the number of days of the year starting from January 1st. **Figure 1** presents the variation of H_0 for Bamenda (latitude 5.96°N and longitude 10.15°E).

The solar declination (δ), the mean sunshine hour angle for the month (ω_s) and the maximum possible sunshine duration (S_0) may be calculated from the Cooper [16] formula [13, 16, 17]:

$$\delta = 23.45 \, sin \left(\frac{360 \, (n_{day} + 284)}{365} \right) \tag{7}$$

$$\omega_s = cos^{-1}(-tan\delta \, tan\varphi) \tag{8}$$

$$S_0 = \frac{2}{15}\omega_s \tag{9}$$

A calculation of these parameters for Bamenda (latitude 5.96°N and longitude 10.15°E) is presented in **Table 1** below.

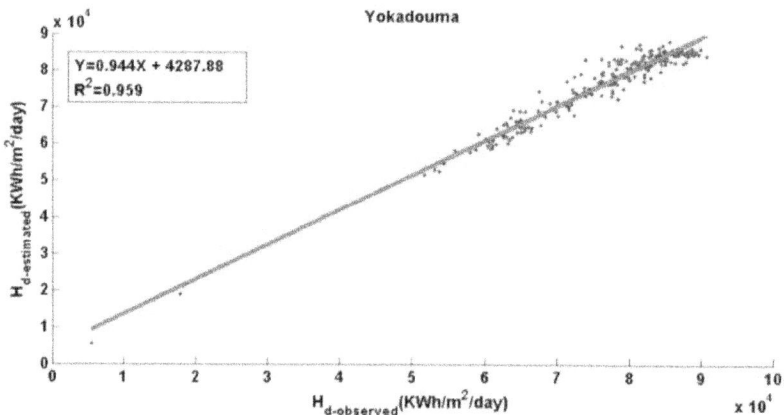

Figure 1.
Correlation between the estimated and observed values of the monthly mean daily diffuse solar radiation using a twenty-year (1985–2005) monthly mean daily clearance index for the area of Yokadouma, Cameroon [15].

Parameter	Value
Average sunshine hours per day	6.7 hours
Solar constant, I_{sc}	1367Wm^{-2}
Latitude of site, φ	5.96^0N
Longitude of site	$10.15°\text{E}$
Linear regression Constants a and b	0.19 and 0.52
Declination, δ from (Eq. (7))	23.18^0
Sunshine hour angle, ω_s from (Eq. (8))	92.56^0
Day length (mean sunshine hour), from (Eq. (9))	12.34 hours
H_0 from (Eq. (5))	$35.6068 \text{ MJ/m}^2/\text{day}$
The estimated value of global solar radiation, H, for Bamenda for the month of June 2005	$16.7756 \text{ MJ/m}^2/\text{day}$
The measured value of global solar radiation, H, for Bamenda for the month of June 2005	$16.452 \text{ MJ/m}^2/\text{day}$

Table 1.
Solar radiation parameters for Bamenda, Cameroon [10].

3.2 Modelling of the direct and diffuse components of solar radiation from GSR measurements, a.k.a. decomposition models

The input parameters for these models are diffuse ratio (K), the clearness index (K_t), the diffuse transmittance index (K_d), and the beam transmittance index (K_b). These parameters are expressed as follows: [18]

$$K = \frac{H_d}{H} \qquad (10)$$

$$K_t = \frac{H}{H_0} \qquad (11)$$

$$K_d = \frac{H_d}{H_0} \qquad (12)$$

$$K_b = \frac{H_b}{H_0} \qquad (13)$$

Where: H is the monthly average daily GSR, H_d is the Monthly average daily diffuse component of GSR, H_b is the Monthly average daily direct component of GSR, H_0 is the Monthly average daily extraterrestrial radiation; H_d is the monthly average daily diffuse radiation received on a horizontal surface, H is the monthly average daily total (direct plus diffuse) radiation received on a horizontal surface, and H_o is the extraterrestrial daily insolation received on a horizontal surface.

3.2.1 Models based on the diffuse ratio- clearness index regressions

The diffuse component of SR can be predicted using GSR data as initially done by Liu and Jordan [19]. The time scales used in this class of models range from monthly average to daily and hourly averages. For monthly average SR, John K. Page [2], estimated the monthly mean values of daily total short wave radiation on vertical and inclined surfaces from sunshine records for latitudes 40° N - 40° S. It consisted of a linear model relating K and K_t. Other similar models have been developed relating K and K_t ranging from quadratic to higher-order polynomial models [20].

For the choice of time scale, some models relate the daily clearness index and daily diffuse SR ratio, for different geographical locations [18, 21, 22]. The approach here consists of developing a piece-wise fit between K and K_t. This is done for overcast, partly-cloudy and clear skies.

For overcast skies, the regression equation is linear and expressed as.

$$K = a_0 + a_1 K_t \text{ for } K_t < K_{ta} \tag{14}$$

Where K_{ta} is some critical value beyond which partly-cloudy conditions dominate.

Other models assume a constant value of K in the event of overcast skies, i.e.:

$$K = a_0 \text{ for } K_t < K_{ta} \tag{15}$$

In the situation of partly cloudy skies, a polynomial fit in K_t of order three or four is used, expressed as:

$$K = b_0 + b_1 K_t + b_2 K_t^2 + b_3 K_t^3 \text{ for } K_{ta} < K_t < K_{tb} \tag{16}$$

Lastly, for a clear sky situation, K takes a constant value, expressed as:

$$K = c_0 \text{ for } K_t > K_{tb} \tag{17}$$

Instead of the piecewise regression as outlined above, a single polynomial regression equation can be chosen that can adequately fit the available data. A nonlinear empirical expression has also been used [23], and given as:

$$K = a + (1 - a) \exp\left[-b K_t^c / (1 - K_t)\right] \tag{18}$$

where, for the location of Macerata the constants take values: a = 0.154, b = 1.062 and c = 0.861.

It should be mentioned that seasonal models in which seasonal variations for daily regressions are treated exist [18, 20, 22].

The third variant consists of the models based on hourly SR measurements. Here the procedure of Liu and Jordan [19], is used with the exception that the correlation between K and K_t is done on an hourly basis [20].

For the performance of these models, **Figure 1** presents the correlation between the estimated and observed values of the monthly mean daily diffuse solar radiation using a twenty-year (1985–2005) monthly mean daily clearance index for the area of Yokadouma, Cameroon, (Latitude 3.15°, longitude 15.050° and at an altitude of 488 m) [15]. For this caption, the correlation equations are expressed in the linear and quadratic forms as: ($H_d = H(1.265 - 1.463K_t)$) and $H_d = H(1.282 - 1.53K_t + 0.063K_t^2)$) [15].

Figure 1 demonstrates a coefficient of determination between the estimated and the observed values close to one (0.92–0.99), which indicates an excellent agreement between the estimated and the observed diffuse fraction. **Figure 2** further shows the correlation between the estimated and observed values of the diffuse fraction for the same location; Yokadouma. Even though the results of the different models follow the same trend, we notice that the predictions of Lealea T. et al. [15] are closest to the observed data. This suggests that these models are location-dependent, performing well in some locations and not in others.

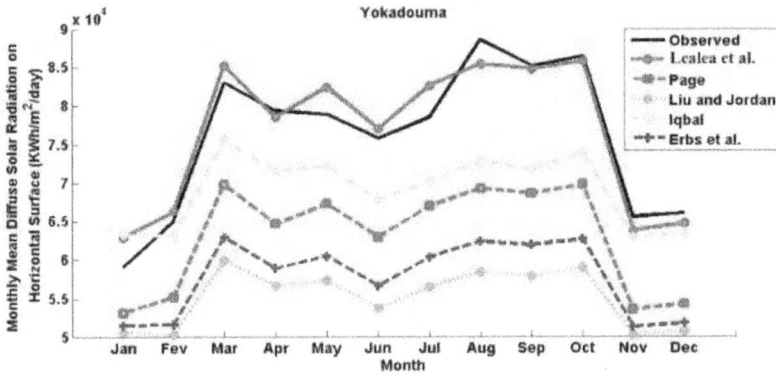

Figure 2.
Comparison of the observed and estimated values of monthly mean diffuse solar radiation predicted by some existing models for Yokadouma, (Cameroon). Modified from [15].

Reference	MBE (Wh/m²/day)	RMSE (Wh/m²/day)	t-statistics
Lealea T. et al. [15] (Linear)	37.8	2615.4	0.228
Lealea T. et al. [15] (Quadratic)	−29.73	2613	0.18
J.K. Page [2]	−13,781	14,261	59.37
Liu and Jordan [24]	−21,751	22,612	55.63
Iqbal [12]	−8092	9824	22.91
Erbs et al. [20]	−19,039	19,737	57.84

Table 2.
Performance of the diffuse ratio- clearness index models using the statistical indicators for Yokadouma, Cameroon [15].

An evaluation of these models based on the statistical indicators for Yokadouma, (Cameroon) is presented in **Table 2**.

The RMSE values here reveal that the model of Lealea T. et al. [15] is best for short-term performance. Meanwhile, the MBE and the RMSE cannot adequately account for the validity of a model, the t-statistics evaluation here indicates that the results of Lealea T. et al. [15] are the most statistically significant for the study location.

3.2.2 Correlation between diffuse transmittance index and clearness index regression

The hourly diffuse SR was predicted from measured hourly GSR on a horizontal surface by Iqbal [25]. It consisted of a correlation between the hourly diffuse transmittance index, k_d, (ratio of diffuse to extraterrestrial radiation), and hourly clearness index, k, (ratio of global to extraterrestrial radiation). The results indicated that the models depend on particular geographical sites.

3.2.3 Correlation between direct transmittance index and clearness index regressions

This approach was spearheaded by Maxwell [26] in an attempt to improve the findings of several investigations which have shown that the use of a single regression function does not sufficiently portray the connection between direct beam transmittance (K_b) and the actual global horizontal transmittance (K_t). The Direct

Insolation Simulation Code (DISC) that uses an exponential relationship between K_b and air mass with parameter K_t, was developed [26]. This procedure acceptably relates K_b with K_t for a variety of stations around the globe and seasons. The validation of the DISC Model exhibited considerable improvements in the correctness of hourly values, substantial decreases in monthly RMS errors, and the corresponding monthly MBE. Further modification of the DISC model integrated the effects of cloud-cover, water vapour, and albedo. Perez et al. [27] adopted and improved on Maxwell's model. The method comprised primarily in using a zenith-angle independent clearness index and by employing a time-varying GSR. Consequently, the diffuse irradiance can be gotten from the difference between the GSR and the beam component once the beam component is known. These components are related by the equation:

$$H = H_d + H_b \, \sin \, \theta_h \tag{19}$$

Where θ_h is the solar height or solar altitude.

Hence Perez et al. [27] improved on the two main shortcomings related to the reliance of the clearness index on solar height and also its slow response to sudden changes in hourly sky conditions.

3.3 Prediction of diffuse solar radiation from the beam or direct component of solar radiation

3.3.1 ASHRAE model

The ASHRAE model considers only clear cloudless days [28]. It proceeds in two steps: the first consists of calculating the intensity of the direct normal solar radiation component and next it computes the hourly direct and diffuse solar radiation on both the horizontal and slanted surfaces. The model equations are:

$$H_{bn} = A \exp(-B/\cos \, \theta_z) \tag{20}$$

$$H_d = CH_{bn} \tag{21}$$

where H_{bn} is the normal beam component of SR, H_d is the diffuse component of SR, θ_z is the zenith angle and A, B, C are monthly mean values of empirically chosen constants.

Extensions of the ASHRAE model where the model coefficients were re-determined using cloudless data at different locations exist [29].

3.3.2 Regression models using the direct transmittance index and the diffuse transmittance index

These models are formulated using an empirical monthly regression equation between the ratio of the daily diffuse SR to the daily extraterrestrial radiation (K_d) and the ratio of the daily beam SR to the daily extraterrestrial radiation (K_b). An implementation in the localities of Beer Sheva and Sde Boker (Israel), is expressed as [30]:

$$K_d = a\left[\exp\left(bK_b + cK_b^2\right)\right] \tag{22}$$

where, the constants a, band c, are monthly values of empirically determined coefficients. For the month of January at Beer Sheva: a = 0.2155, b = 3.1713 and c = −8.1261.

3.4 Prediction of solar radiation from meteorological input parameters

These models are formulated using the clearness index, k_t with the input parameter being the GSR. They present a shortcoming in that the clearness index alone cannot account for changes in the diffuse component of the SR. Extensions of the model exist that can address the associated drawback that will be explored in the following subsections.

3.4.1 Prediction of solar radiation from the sunshine fraction

The first attempt that expresses SR in terms of the sunshine fraction is the linear equation [3]:

$$\frac{H}{H_0} = a + b\left(\frac{S}{S_0}\right) \tag{23}$$

where a and b are the two constants, H is the monthly average daily SR, H_o is the monthly average daily extra-terrestrial radiation, S is the monthly average daily measured sunshine duration.

As an extension of this equation and to improve the accuracy, the nonlinear polynomial models, were derived. This form is given as follows [7]:

$$\frac{H}{H_0} = a + b\left(\frac{S}{S_0}\right) + c\left(\frac{S}{S_0}\right)^2 + d\left(\frac{S}{S_0}\right)^3 + \dots \tag{24}$$

The values of a, b, c and d, vary depending on location and month of observation. Their values may be affected by atmospheric air pollution. As the daily total amount of SR and sunshine duration vary widely, daily totals averaged over a month are used to derive the values of a, b, c and d. This can be done by the least square regression analysis.

These models have been very popular all the time because of the abundance of data on sunshine duration in most locations on earth. This eases the prediction of GSR even where measurements are absent. The mostly used equation is that proposed by John K. Page [2], expressed as:

$$H = H_0\left(a + b\frac{n}{N}\right) \tag{25}$$

where H and H_0 are the monthly-average daily terrestrial and extraterrestrial radiation, n is the average daily hours of bright sunshine and N is the day length. Variants of these models are linear, quadratic, third-degree polynomial, exponential and logarithmic (**Figures 3** and **4**).

To test these models, we present results for both the linear, the quadratic, and the third-degree polynomial models for the city of Bamenda in Cameroon whose parameters have been presented in **Table 1**.

3.4.2 Cloud cover radiation models (CRM)

For cloud cover radiation models, the choice parameters used are the monthly mean values of the fraction of the sky covered by clouds, $\overline{Ne}$, and duration of bright sunshine hours, N. The sunshine duration is calculated from the cloud cover data and the cloud derived sunshine data, monthly mean values of global and diffuse SR are calculated. The model equations are expressed as [31]:

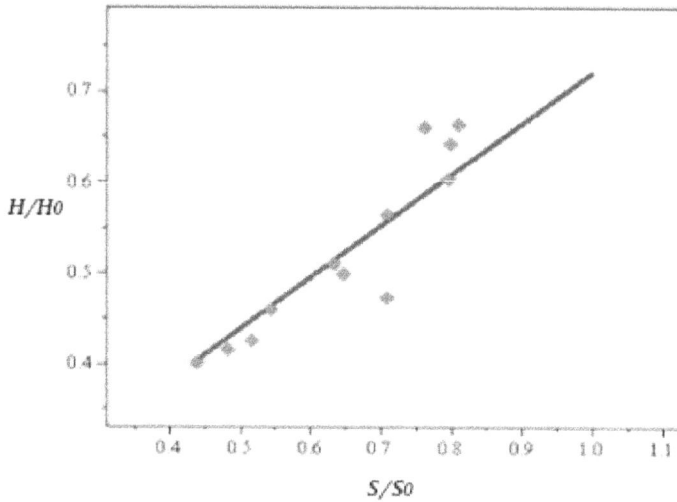

Figure 3.
Linear relationships between the monthly average values (H/H_o versus S/S_o) for the city of Bamenda, Cameroon [10].

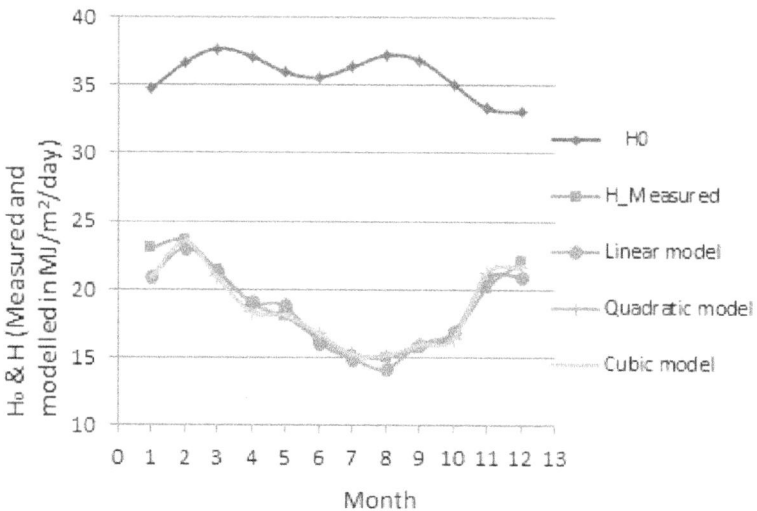

Figure 4.
Comparison of the estimated and observed monthly average daily horizontal GSR data for Bamenda using the linear, quadratic and cubic models.

$$G_{t,0} = W \sin \theta_h - X \qquad (26)$$

$$\frac{G_t}{G_{t,0}} = 1 - Y \left(\frac{N_e}{8}\right)^Z \qquad (27)$$

Where, $G_{t,0}$ is the global irradiance for a cloudless sky in W/m^2, G_t is the hourly global irradiance for any given cloud amount (N_e, in eighths) in W/m^2, θ_h is the solar height in (°), and W, X, Y and Z are empirical regression constants.

A linear model equation that correlates monthly average diffuse transmittance index, $\overline{K}_d$, to monthly average daily cloud cover ($\overline{N}_e$, in eighths), is given by [32]:

$$\overline{K}_d = a_7 + b_7 \frac{\overline{N}_e}{8} \qquad (28)$$

$$\overline{K}_d = a_8 + b_8 \left(1 - \frac{\overline{N}_e}{8}\right) \qquad (29)$$

Where a_7, b_7, a_8 and b_8 are regression constants.

As an extension, several empirical models for the prediction of GSR from the daily mean of cloud cover, temperature extremes (minimum and maximum) and extraterrestrial SR have been proposed [33].

3.4.3 Models based on atmospheric transmittance (ATM)

Constituents that affect the transmittance of the atmosphere include scatterers, which consist of air molecules responsible for Rayleigh scattering, aerosols causing Mie scattering and absorbers like water vapour, atmospheric gases, dust and clouds. The atmospheric transmittance models attempt to establish some parametric relationships between these parameters. These models can either be classified as broadband or spectral based/ radiative transfer models [34].

3.4.3.1 Meteorological radiation model (MRM)

The most popular broadband ATM is the Meteorological Radiation Model (MRM). The input data for this model consist of the dry- and wet- bulb temperature and a sunshine fraction (used to generate hourly SR components for all-sky conditions like overcast and clear skies) [31].

The model equations for MRM in the case of non-overcast skies are given by

$$DBR = 0.285k_b^{-1.00648} \qquad (30)$$

$$G_b = (SF)G_0\tau_r\tau_\alpha\tau_g\tau_o\tau_w \qquad (31)$$

Where: DBR is the hourly diffuse to beam ratio, k_b is the direct transmittance index, G_b is the beam/direct irradiance, SF is the hourly sunshine fraction, τ_r, τ_α, τ_g, τ_o, τ_w are the transmittances respectively due to Rayleigh and Mie scattering, mixed gases, ozone and water vapour. Empirical equations are used to determine the transmittance indices and the coefficients are obtained through data fitting.

The proposed model exists for overcast skies where the diffuse irradiance is assumed to be equal to GSR [35]. Gueymard [36] proposed another similar model referred to as the Reference Evaluation of Solar Transmittance (REST). Though similar to the other models, the particularity of REST is that it introduces an additional transmittance term τ_n to account for the total absorption of NO_2.

3.4.3.2 Spectral models

The measurement of the solar spectrum is quite challenging necessitating models that can accurately provide the solar radiation incident at different parts of the earth's surface.

Spectral models are particularly suitable for such applications that are prone to small changes in wavelength. These models are spurred on the one hand by the challenges encountered in measurements of the electromagnetic spectrum. On the other hand, there is a need for models capable of accurately reproducing the incident radiation at the earth's surface. This aim is achieved by solving the radiative transfer equations as a function of the wavelength intervals as well as unit atmospheric layer intervals [37, 38]. The first of these models are the SPECTRAL and the Simple Model of the Atmospheric Radiative Transfer of Sunshine (SMARTS) developed by Bird [37]. The second is a modified version of SPECTRAL to SPECTRAL2 developed eventually by Bird and by Riordan [38]. These models apply simple mathematical expressions on tabulated look-up tables to generate the direct-normal and diffuse horizontal irradiance.

The SPECTRAL2 determines the beam component of solar radiation perpendicular to the earth surface for some wavelength λ through the equation:

$$H_{b\lambda} = H_{o\lambda} D T_{r\lambda} T_{a\lambda} T_{w\lambda} T_{o\lambda} T_{u\lambda} \tag{32}$$

Where for some given wavelength λ and for some mean earth-sun distance: $H_{o\lambda}$ is the extraterrestrial irradiance; D is a correction factor; and $T_{r\lambda}$, $T_{a\lambda}$, $T_{w\lambda}$, $T_{o\lambda}$, and $T_{u\lambda}$ are functions expressing the transmittance of the atmosphere for molecular Rayleigh scattering, attenuation by aerosols, absorption by water vapour, absorption by ozone, and absorption by uniformly mixed gases, respectively. The beam component of solar irradiation on a horizontal surface is given by the product of (Eq. (32)) and the cosine of the solar zenith angle, θ_z.

The parameter, D in (Eq. (32)) is expressed as

$$D = 1.00011 + 0.034221 \; cos \; \omega_d + 0.00128 \; sin \; \omega_d + 0.000719 \; cos \; 2\omega_d$$
$$+ 0.000077 \; sin \; 2\omega_d \tag{33}$$

Where ω_d the day angle in radians given by

$$\omega_d = 2n \; (n_{day} - 1)/365 \tag{34}$$

Three components make up diffuse solar radiation on a horizontal surface. The first $H_{r\lambda}$ results from the Rayleigh scattering, the second $H_{a\lambda}$ is caused by the aerosol scattering and the third $H_{g\lambda}$ originates from multiple reflections of solar radiation between the earth surface and the atmosphere. The resultant solar radiation caused by scattering is expressed as:

$$H_{s\lambda} = H_{r\lambda} + H_{a\lambda} + H_{g\lambda} \tag{35}$$

Obtaining the spectral solar radiation on inclined surfaces is a straight forward process achieved by combining the spectral beam and diffuse radiation components calculations as presented above. The spectral global solar radiation on a slanted surface is then given by

$$H_{T\lambda}(t) = H_{b\lambda} \cos \theta + H_{s\lambda}[(H_{b\lambda} \cos \theta/H_{o\lambda} D \; \cos \theta_z) + 0.5(1 + \; cos \; \beta)(1 - H_{b\lambda}/(H_{o\lambda} D))]$$
$$+ 0.5 H_{T\lambda} r_{g\lambda}(1 + \; cos \; \beta) \tag{36}$$

where θ is the angle of incidence of the beam component on the tilted surface, β is the angle of tilt of the slanted surface relative to the horizontal surface and θ_z is the solar zenith angle. The following expression holds for the spectral global solar radiation on a horizontal surface:

$$H_{T\lambda} = H_{b\lambda} \cos \theta_z + H_{s\lambda} \tag{37}$$

The first term in (Eq. (36)), is the direct component on the inclined surface. The second term has two components: the first is circumsolar and the second is a diffuse

component. The last term represents the radiation reflected from the earth surface
which is distributed isotropically. A component that is missing from this model is
the horizon-brightening radiation.

Gueymard [39] improved the SMARTS model to the SMARTS2. The spectral
transmittance is expressed as a function of the processes responsible for radiation
extinction in the atmosphere such as water vapour, Rayleigh scattering, uniformly
mixed gases, absorption by ozone, aerosol extinction and Nitrogen dioxide. These
functions are then used to calculate the beam component of the radiation in the
shortwave range. Data obtained from spectroscopic measurements have been used
to calculate coefficients for the extinction processes due to absorption by gases that
depend on both temperature and pressure. The coefficient of absorption resulting
from the dependence in temperature is captured in the modelling of the extinction
caused by nitrogen dioxide, both in the visible and UV regions of the electromagnetic spectrum. The two-tier Angstrom methodology is used to compute the
extinction resulting from absorption by aerosols. Data of visibility measured at the
airport and further refined based on a prototype of the Shettle and Fenn [40]
function is used to evaluate the turbidity effect of aerosols. A further improvement
is introduced by expressing the wavelength exponent and some coefficients that
characterise the individual aerosol components as a multivariable parametric function of the relative humidity and the wavelength. SMARTS2 is also equipped with
an optional function that corrects the circumsolar radiation which together with
two functions that smoothen and filter the spectral solar radiation equip it with the
possibility to mimic real-time spectro-radiometers. As a result, confronting the
results of modelling with observed data becomes easy. An initial evaluation of the
validity of SMARTS2 revealed considerable agreement for the direct component of
solar radiation obtained both from thorough and standard solar radiation schemes
and from spectro-radiometric measurements. The possibility of incorporating into
SMARTS2 the ability to estimate solar spectra under the canopy of clouds is further
suggested in a later work by Gueymard et al. [41].

3.5 Satellite-based models

Geographical and climate parameters vary widely across the globe and consequently impute differences in the amount of SR intercepting the earth's surface. To
capture all these differences would require an infinite number of ground measuring
stations. This difficulty is alleviated by the use of meteorological satellites which
provide SR data over a wide geographical coverage with high spatial resolution.
Models based on such data have been developed to take advantage of such ubiquitous data. The models range from: subjective, empirical (statistical and physical
based), objective and theoretical (broadband and spectral) [42].

3.5.1 Subjective methods

Methods that involve some subjective interpretation of the satellite data fall under
this category. For the method to provide some quantitative measure for solar radiation,
it has to be associated with other methods. This method has been applied manually to
estimate cloud cover from hard-copy images using a gridded overlay [43].

3.5.2 Empirical methods

Here functional relations are developed using simultaneous and co-located satellite and SR data. The methods permit some level of transferability in which the
derived equations can be applied to other geographical locations, but as pointed out

in [42], such a process may be uncertain due to the empirical parameters involved in the equations. In the subsequent development of these methods, two approaches are followed: the first is a statistical approach and the second is a physical approach.

The statistical approach relies upon choosing the independent variable merely based on their facility to capture the trend in the SR based on the geographical location of interest. In what follows, the physical-based approach will be prioritized and developed.

3.5.2.1 Physical based methods

These methods originate from an attempt to achieve a radiation balance between the earth and its surrounding atmosphere. A formulation presented in [42], expresses the balance as follows:

$$E_0 \downarrow - E_0 \uparrow - E_a - E_g \downarrow (1 - \rho) = 0 \tag{38}$$

Where $E_0 \downarrow$ is the extraterrestrial solar irradiance, $E_0 \uparrow$ is the SR reflected back to space, E_a is the SR absorbed by the earth's atmosphere, $E_g \downarrow$ is the solar irradiance at the earth's surface, and ρ is the surface albedo.

Dividing by $E_0 \downarrow$ and rearranging terms results in:

$$\rho_p = 1 - q_a - q_t(1 - \rho) \tag{39}$$

where ρ_p is the planetary albedo (the fraction of the incident SR reflected to space); q_a is the portion of the incident SR absorbed by the atmospheric constituents; q_t is the transmitted portion of the incident SR through the atmosphere. Using an argument whereby q_t and the spatially averaged values of ρ_p are highly correlated, enables the use here of a statistical equation of the form

$$\rho_p = a + bq_a \tag{40}$$

where a and b are some empirical coefficients equal to 0.63 and -0.64, respectively [44].

It can be deduced from (Eq. (39)) that,

$$a = 1 - q_a \tag{41}$$

and

$$b = -(1 - \rho) \tag{42}$$

This results in average values of 0.37 and 0.36 for q_a and ρ, respectively. However, it was shown in [44] that the satellite data could have undervalued ρ_p resulting in an overestimation of the atmospheric absorption values inferred.

Eq. (39) can be alternatively expressed in the form

$$q_t = \left(1 - \rho_p - q_a\right)/(1 - \rho) \tag{43}$$

If all the quantities in this equation are obtained from appropriate measurements, then q_t can be calculated.

An alternative approach was followed in [45] to develop a model in which there is a very high correlation between the planetary albedo and the SR absorbed at the earth's surface, thereby implying that the column integrated atmospheric absorption is highly conservative. On this basis, the model is expressed as:

$$q_t(1 - \rho) = a + b\rho_p \tag{44}$$

Where it can be deduced from equation (Eq. (39)), that:

$$a = 1 - q_a \tag{45}$$

and

$$b = -1 \tag{46}$$

The conservative aspect of the regression parameters was revealed by using data from different geographical locations. This was further substantiated theoretically leading to the conclusion that even clouds cannot severely change the atmospheric absorption.

Other empirical models have been developed based on the radiation balance between the earth and its surrounding atmosphere [46, 47]. One approach followed in [48] and [49] consists of rearranging Eq. (39) in the form:

$$q_t = \left(1 - \rho_p - q_a\right)/(1 - \rho) \tag{47}$$

This equation can be rewritten in the form:

$$q_t = a + b\rho_p \tag{48}$$

Where,

$$a = \left(1 - q_a\right)/(1 - \rho) \tag{49}$$

and

$$b = -1/(1 - \rho) \tag{50}$$

A comprehensive analysis in [50] and [51] led to expressing the parameters, a and b as multivariable functions given by:

$$a = f\left(q_a, \rho_p, \rho'_p, \rho'_c, \rho'\right) \tag{51}$$

$$b = f\left(q_a, q'_a\rho_p, \rho'_p, \rho_c, \rho'_c, \rho, \rho'\right) \tag{52}$$

where: ρ_c is cloud reflectivity and the primes indicate that the variable is calculated when the satellite sensor is in some spectral interval (typically 0.55–0.75 μm). They revealed that b is less conservative than a, as a consequence of the relatively strong reliance on aerosol absorptivity which is one component of q_a and q'_a. The other two most important parameters (cloud reflectivity and water vapour absorptivity) neutralize each other. Hence, the combined effect of these latter variables may be very inconsequential.

Cano et al. [52] developed a model that deviates from the previous ones and can serve as a transition between the empirical and theoretical models. In their approach, the cloudless sky albedo is computed iteratively by a procedure that minimizes the variance in the difference between the satellite inferred value of $E_g \uparrow$ and a calculated value obtained from:

$$E_g \uparrow = \rho_{po} I_T (\cos \theta_z)^{1.15} \tag{53}$$

In this approach, ρ_{po}, is the planetary albedo for a cloudless target. I_T is the solar constant corrected for actual sun-earth distance. A cloud cover index (n_s) is computed for high surface albedo (e.g., with snow cover) and for infrared radiances for wavelengths in the interval between 10.5 and 12.5 μm as

$$n_s = \frac{I - I_0}{I_c - I_0} \tag{54}$$

Where I is the observed infrared radiance, I_0 is the observed infrared radiance for cloudless sky conditions and.

I_c is the observed infrared radiance for overcast sky conditions.

The atmospheric transmission was assumed to be a linear combination of the respective values for cloudless and overcast skies resulting in

$$q_t = (1 - n_s)q_{t0} + n_s q_{tc} = q_{t0} + n_s(q_{tc} - q_{t0}) \tag{55}$$

In a similar regression model

$$q_t = a + b \, n_s \tag{56}$$

with

$$a = q_{t0} \tag{57}$$

and

$$b = q_{tc} - q_{t0} \tag{58}$$

According to Cano et al. [52], if data were stratified hourly, absolute values of the correlation coefficients are typically greater than 0.80, thereby supporting their use of the preceding model. Although the parameters, a and b could be calculated analytically, values were determined empirically. This is in line with the fact that the regression parameters also account for many other effects, including those resulting from the characteristic response of the satellite sensors.

3.5.3 Theoretical methods

These models endeavour to simulate explicitly solar radiant energy exchanges occurring between the earth and the atmosphere. Unlike the statistical counterpart, they do no incorporate an empirical calibration of the model parameters resulting in location free models. The models however need to be provided with additional environmental data which are time and location dependent. As a short cut to this limitation, climatological and standard atmosphere data are sometimes used, often without seriously impacting negatively on model performance.

Based on the degree of simplification and realism, two general classes of models can be distinguished: broadband models formulated based on the earth's radiation balance and spectral models which rely on results generated by the solution of the radiative transfer equation.

3.5.3.1 Broadband models

One of the pioneers in this approach is Gautier et al. [53], who developed a model that has been widely used and makes it a reference for broadband models. In their model, the solar flux that exits the earth's atmosphere and is measured by the satellite is given by:

$$E_0 \uparrow = E_0 \downarrow - E_a - E_g \downarrow (1 - \rho) \tag{59}$$

By starting with a cloudless sky and minimizing the effects of multiple reflections down to first-order and assuming that scattering occurs before absorption, Gautier et al. rewrote Eq. (59) in terms of broadband absorption and scattering coefficients to give

$$E_0 \uparrow = E_0 \downarrow (\rho' + (1 - \rho')(1 - a_1)(1 - a_2)(1 - \rho^* \rho)) \tag{60}$$

The only unknown is the surface albedo, ρ. $E_0 \uparrow$ can be inferred from the satellite measurements, ρ' and ρ^* are obtainable from Coulson [42] and a_1 and a_2 can be calculated given an estimate of atmospheric water vapour content (commonly climatological data or relationships involving surface humidity are used). Making ρ the subject in the last equation gives:

$$\rho = \frac{E_0 \uparrow - E_0 \downarrow}{E_0 \downarrow (1 - \rho')(1 - a_1)(1 - a_2)(1 - \rho^*)} \tag{61}$$

A rearranged and expanded version of (Eq. (60)) was then used to express the solar irradiance at the earth's surface (assuming cloudless skies) in terms of known variables

$$E_g \downarrow_0 = E_0 \downarrow (1 - \rho')(1 - a_1)(1 - a_2)(1 + \rho^* \rho) \tag{62}$$

This model was revised by Diak and Gautier [54], where they included the effects of ozone absorption while Gautier and Frouin [55] also incorporated the effects of both aerosol and all orders of multiple reflections. Additional revisions investigated the consequences of ignoring spectral dependencies in both atmospheric attenuation and satellite radiometers.

Diak and Gautier [54] recognized that: (1) the Rayleigh scattering optical depth is wavelength dependent and therefore values of ρ' and ρ^* must be evaluated for both the entire solar spectrum and the wavelengths covered by the visible sensors. (2) Ozone absorption must be considered, especially given its significance in the visible part of the spectrum. These considerations are captured in the calculations of both the albedos and surface irradiances. Gautier and Frouin [55] provided the following equation for the surface irradiance during cloudless skies:

$$E_g \downarrow_0 = E_0 \downarrow \exp\left(-\frac{c_2}{\cos \theta_z}\right)(1 - a_{01})(1 - a_{03})(1 - a_1)/(1 - c_3 \rho) \tag{63}$$

Gautier et al. [53] revised and extended their model to include the effects of clouds by assuming a plane-parallel atmosphere composed of three layers. Thirty per cent of the water vapour and all the Rayleigh scattering were assumed confined to the top cloud layer. Similar procedures to those considered in the clear sky model provided estimates of the cloud top albedo (ρ_0) with which the cloud absorption (a_0) was parameterized using a linear function ranging from zero for no cloud to 20% absorption for maximum target brightness.

Multiple reflections were not considered in the derivation of the following equation for the irradiance at the surface under overcast conditions [53]

$$E_g \downarrow_c = E_0 \downarrow (1 - \rho')(1 - a_{1a})(1 - \rho_c)(1 - a_c)(1 - a_{1b}) \tag{64}$$

We notice a striking similarity with (Eq. (62)) (for clear skies) except that the first order of multiple reflections was included in that formulation. Note also that to be consistent with the definition of ρ_c and a_c, the term $(1 - \rho_c)(1 - a_c)$ should be replaced by $(1 - \rho_c - a_c)$.

A revised equation for $E_g \downarrow_c$ [54] captured the effects of ozone – zone absorption and revised the formulation of cloud attenuation to render it consistent with the definition of the absorption and reflection coefficients

$$E_g \downarrow_c = E_0 \downarrow (1 - a_{01})(1 - a_{03})(1 - \rho')(1 - a_{1a})(1 - \rho_c - a_c)(1 - a_{1b}) \tag{65}$$

Further attempts were made to approach reality in the parameterization of absorption by cloud, primarily to incorporate the occurrence of both finite and sub-field-of-view clouds. A decision to limit ρ_c to values greater than 7% was arrived at based on comparisons between measured and calculated surface irradiances. For analogous reasons, the maximum value of a_c was set to 7%. The basis for these decisions to an extent weakens the claim of zero empiricism in physically-based models.

Gautier and Frouin [55] upgraded their analysis to capture the effects of both aerosols and multiple reflections resulting in

$$E_0 \downarrow_c = c_1 E_0 \downarrow exp\left(-\frac{c_2}{\cos\theta_z}\right)(1 - \rho_c - a_c)(1 - a_{01})(1 - a_{03})(1 - a_1)/(1 - c_3\rho)(1 - c_4\rho) \tag{66}$$

Gautier et al. [53] have described the procedures for deciding whether to implement the clear or overcast sky routines when calculating $E_g \downarrow$, for a given location (pixel) and the technique for combining these estimates for partly cloudy conditions.

From the foundational investigations of Gautier et al. [53], other works have followed and revised their models to address some of the shortcomings associated with their approach such as the investigations in [56].

3.5.3.2 Spectral models

The conceptual basis for modelling SR by a spectral model of radiative transfer is best captured in the technique developed by Halpern [57]. The solution of the radiative transfer equation for an atmosphere tending to be absorbing and scattering requires some simplifying assumptions. Dave and Braslau [58] used a direct numerical solution of the spherical harmonics approximation for the axially symmetric but highly anisotropic phase functions which describe the scattering properties of liquid water drops (cloud) and aerosol. Halpern [57] used Dave and Braslau [58] results to construct tables of the downward ground-level flux and the upward flux at the top of the atmosphere. These initial attempts have been refined in two main aspects. The limitations imposed by the discrete nature of the Halpern approach are avoided, through the use of parameterizations based on data provided by explicit solutions of the radiative transfer equation for a wide range of atmospheric conditions. The algorithms are typically independent of conventional data sources, with all site and time-specific environmental data being abstracted from the digital imagery.

Moser and Raschke [59] also using a radiative transfer model developed the following parameterizations for several model atmospheres and a wide range of boundary conditions

$$E_g \downarrow_0 = f(\theta_z) \tag{67}$$

$$E_0 \uparrow_c = f(\theta_z, h_c) \tag{68}$$

$$\frac{E_g \downarrow}{E_g \downarrow_0} = 1 - f(\theta_z, L_n) \tag{69}$$

The cloud height (h_c) was determined using simultaneous satellite measurements in the infrared (10.5–12.5) μm while L_n (the normalized reflected radiance) was determined from

$$E_g \downarrow = E_0 \uparrow [1 - f(\theta_z, L_n)] \tag{70}$$

3.6 Classification and comparative study of the models

We summarise here the models presented in the previous sections aiming to show the interrelationship amongst the models and the input and output parameters of each (**Table 3**).

3.6.1 Classification of the models

See **Figure 5**.

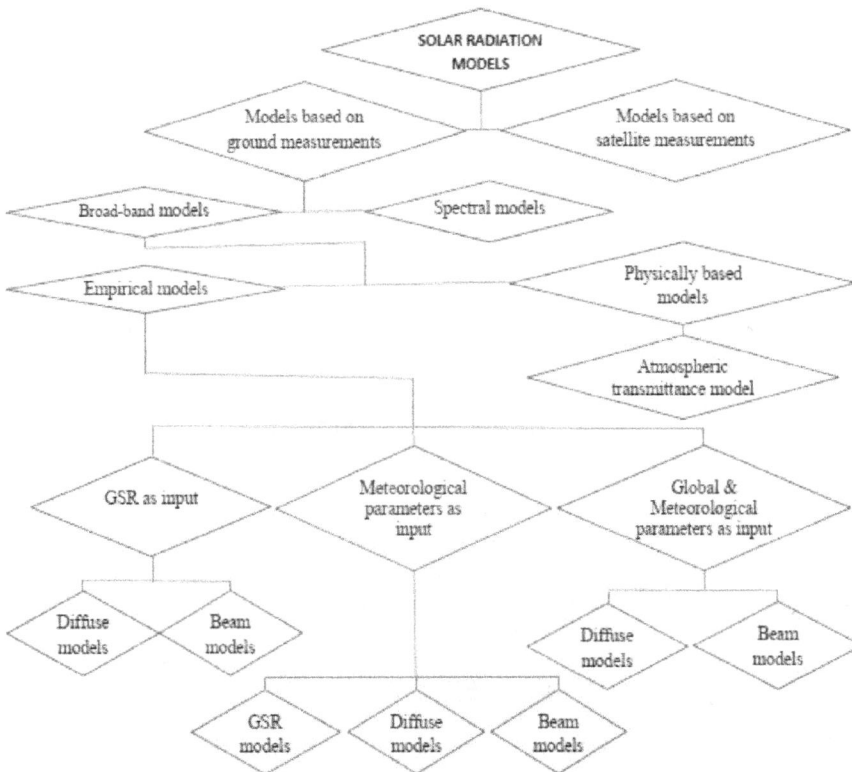

Figure 5.
Classification of solar radiation models. Modified from [60].

3.6.2 Comparative study of the models

Model	References	Input parameters	Output parameters	Linearity
Models based on the diffuse ratio- clearness index regressions	[2, 15, 18–20]	Clearness index (K_t)	Diffuse ratio (K)	Linear and Nonlinear
Models based on diffuse transmittance index and clearness index regression	[25]	Hourly clearness index, k	Hourly diffuse transmittance index, k_d	Linear
Models based on direct transmittance index and clearness index	[26, 27].	Clearness Index (K_t)	Direct beam transmittance (K_b), air mass, cloud-cover, water vapour, and albedo	Nonlinear
ASHRAE model	[28, 29]	Zenith angle, θ_z	Diffuse radiation, H_d	Nonlinear
Models using the direct transmittance index and the diffuse transmittance index	[30]	Diffuse transmittance index (K_d)	Beam transmittance index (K_b)	Nonlinear
Models based on the sunshine fraction	[2, 3, 10]	Solar fraction, S/S_0	Clearness Index (K_t)	Linear and Nonlinear
Cloud cover radiation models (CRM)	[32, 33]	Monthly average daily cloud cover ($\overline{N}_e$, in eighths)	Monthly average diffuse transmittance index, $\overline{K}_d = \overline{H}_d/\overline{H}$,	Linear
Meteorological Radiation Model (MRM)	[31, 35, 36]	SF, τ_r, τ_α, τ_g, τ_o, τ_w, k_b	DBR, G_b	Nonlinear

Table 3.
Comparative study of the Solar radiation models.

4. Tilt and azimuth angles in solar photovoltaics energy applications

4.1 Introduction

The aims in this section is to present the optimum tilt angles calculation methods required for the optimal and best design of solar PV systems. Some techniques applicable for solar tilt calculations have been elaborated in [61, 62]. Some valuable excerpts from these references are considered in this section.

4.2 Optimal tilt angles for global solar radiation components

Like on horizontal surfaces, the total daily radiation falling on tilted surfaces (G_T) is the sum of three components: the direct (G_{B_t}), diffuse (G_{D_t}) and ground reflected (G_{R_t}). This is expressed as [62] (**Figure 6**)

$$G_T = G_{B_t} + G_{D_t} + G_{R_t} \tag{71}$$

These three components are respectively related to direct, diffuse and total radiation on horizontal surfaces through the three expressions

$$G_{B_t} = R_b G_B \tag{72}$$

$$G_{D_t} = R_d G_D \tag{73}$$

$$G_{R_t} = R_r G \tag{74}$$

R_b, R_d and R_r are the quotients of the daily solar radiation incident on a slanted surface to that incident on a horizontal surface for the beam, the diffuse and the reflected components respectively. G_B, G_D and G are the beam, diffuse and total daily SR on a horizontal surface. (Eq. (71)) then asumes the expression:

$$G_T = R_b G_B + R_d G_D + R_r G \tag{75}$$

The calculation of the direct and diffuse components of GSR needed for the estimation of GSR on slanted surfaces was well elaborated in subSection 3.1.

In terms of the albedo, ρ, and the tilt angle of the horizontal surface β, R_r is expressed as:

$$R_r = \rho \left(\frac{1 - \cos \beta}{2} \right) \tag{76}$$

Here R_b depends on the transmittance of the atmosphere which is in turn affected by the atmospheric cloud cover, water vapour and concentration of atmospheric particles.

R_b for fixed slope surfaces oriented towards the equator in the northern hemisphere is expressed in [61, 64] as

$$R_b = \frac{\cos(\varphi - \beta) \cos \delta \sin \omega_{st} + \omega_{st} \sin(\varphi - \beta) \sin \delta}{\cos \varphi \cos \delta \sin \omega_{st} + \omega_{st} \sin \varphi \sin \delta} \tag{77}$$

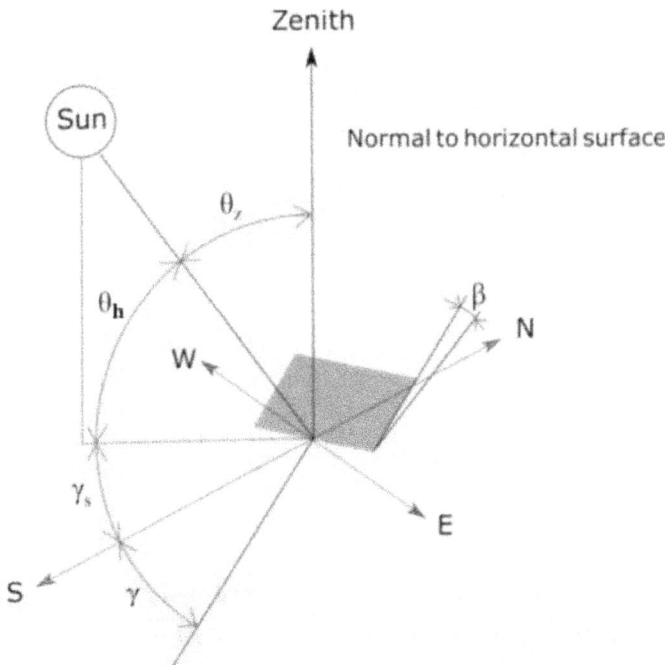

Figure 6.
Zenith angle (θ_z), slope (β), surface azimuth angle (γ) and solar azimuth angle (γs) for a tilted surface. Modified from [63].

Where ω_{st} is the sunset hour angle for the tilted surface, for the mean day of the month, which is given by

$$\omega_{st} = min\left[cos^{-1}(-\tan\varphi\tan\delta), \; cos^{-1}(-\tan(\varphi-\beta)\tan\delta)\right] \tag{78}$$

For surfaces in the southern hemisphere sloped towards the equator, the equations are [61, 65]:

$$R_b = \frac{\cos(\varphi+\beta)\cos\delta\sin\omega_{st} + \omega_{st}\sin(\varphi+\beta)\sin\delta}{\cos\varphi\cos\delta\sin\omega_{st} + \omega_{st}\sin\varphi\sin\delta} \tag{79}$$

$$\omega_{st} = min\left[cos^{-1}(-\tan\varphi\tan\delta), \; cos^{-1}(-\tan(\varphi+\beta)\tan\delta)\right] \tag{80}$$

It is possible to alternatively estimate R_b as the quotient of the daily extraterrestrial radiation on the slanted surface to that on a horizontal surface, G_0 [24]. As a consequence the following relation has been proposed for R_b [62, 66]:

$$
\begin{aligned}
R_b &= \int_{\omega_{rt}}^{\omega_{st}}\cos\theta(\omega)d\omega \Big/ \int_{\omega_r}^{\omega_s}\cos\alpha(\omega)d\omega \\
&= (\cos\beta\sin\delta\sin\phi)\left(\frac{\pi}{180}\right)(\omega_{st}-\omega_{rt}) - (\sin\delta\cos\phi\sin\beta\cos\alpha_t)(\pi/180)(\omega_{st}-\omega_{rt}) \\
&\quad + (\cos\phi\cos\delta\cos\beta)(\sin\omega_{st} - \sin\omega_{rt}) + (\cos\delta\cos\alpha_t\sin\phi\sin\beta)(\sin\omega_{st} - \sin\omega_{rt}) \\
&\quad + (\cos\delta\sin\beta\sin\alpha_t)(\cos\omega_{st} - \cos\omega_{rt})) \Big/ \left(2\left(\cos\phi\cos\delta\sin\omega_s + \left(\frac{\pi}{180}\right)\omega_s\sin\phi\sin\delta\right)\right)
\end{aligned} \tag{81}
$$

Where ω_r and ω_s are the sunrise and sunset hour angles over the horizon respectively in degrees. Also, ω_{rt} and ω_{st} are sunrise and sunset hour angles over tilted plane surface calculated as follows:
if $\alpha_t < 0$,

$$\omega_{rt} = -min\left(\omega_s, \; cos^{-1}\left(\frac{PQ+\sqrt{P^2-Q^2+1}}{P^2+1}\right)\right) \tag{82}$$

$$\omega_{st} = min\left(\omega_s, \; cos^{-1}\left(\frac{PQ-\sqrt{P^2-Q^2+1}}{P^2+1}\right)\right) \tag{83}$$

else

$$\omega_{rt} = -min\left(\omega_s, \; cos^{-1}\left(\frac{PQ-\sqrt{P^2-Q^2+1}}{P^2+1}\right)\right) \tag{84}$$

$$\omega_{st} = min\left(\omega_s, \; cos^{-1}\left(\frac{PQ+\sqrt{P^2-Q^2+1}}{P^2+1}\right)\right) \tag{85}$$

Where,

$$P = \cos\phi/(\sin\alpha_t\tan\beta) + \sin\phi/\tan\alpha_t \tag{86}$$

$$Q = \tan\delta\,(\cos\phi/\tan\alpha_t - \sin\phi/(\sin\alpha_t\tan\beta)) \tag{87}$$

4.3 Diffuse radiation models on tilted surfaces

Both isotropic and anisotropic models exist for estimating the ratio of diffuse SR on a tilted surface to that on a horizontal surface. The isotropic models assume the

Reference	Optimum Tilt Angle expressed in terms of altitude
Isotropic models	
[65]	$R_d = \frac{3 - \cos 2\beta}{4}$
[67]	$R_d = 1 - \frac{\beta}{180}$
[68]	$R_d = \frac{2 + \cos\beta}{3}$
[24]	$R_d = \frac{1 + \cos\beta}{2}$
Anisotropic models	
[69]	$R_d = \frac{G_B}{G_{B_n}} R_B + \left(1 - \frac{G_B}{G_{B_n}}\right)\left(1 + \frac{\cos\beta}{2}\right)\left(1 + \sqrt{\frac{G_B}{G_{B_t}}}\, \sin^3\left(\frac{\beta}{2}\right)\right)$
[70]	$R_d = \frac{G_B}{G_{B_n}} R_B + \Omega\cos\beta + \left(1 - \frac{G_B}{G_{B_n}}\right)\left(1 + \frac{\cos\beta}{2}\right)\left(1 + \sqrt{\frac{G_B}{G_{B_t}}}\sin^3\left(\frac{\beta}{2}\right)\right),$ Where $\Omega = \max\left[0, \left(0.3 - 2\frac{G_B}{G_{B_n}}\right)\right]$
[71]	$R_d = 0.51 R_b + \frac{1 + \cos\beta}{2} - \frac{1.74}{1.26\pi}\left[\sin\beta - \left(\beta\frac{\pi}{180}\right)\cos\beta - \pi\sin^2\left(\frac{\beta}{2}\right)\right]$
[72]	$R_d = \frac{G_B}{G_{B_n}} R_B + \left(1 - \frac{G_B}{G_{B_n}}\right)\left(1 + \frac{\cos\beta}{2}\right)$
[73]	$R_d = \left(\frac{1 + \cos\beta}{2}\right)\left(1 + F\sin^3\left(\frac{\beta}{2}\right)\right)\left(1 + F\cos^2\theta \sin^2\theta_z\right)$ Where, $F = 1 - \left(\frac{G_{D_t}}{G_B}\right)^2$, $\cos\theta = \sin\delta\sin(\varphi - \beta) + \cos\delta\cos(\varphi - \beta)\cos\omega\sin\theta_z$ $= \sqrt{1 - (\sin\delta\sin\varphi + \cos\delta\cos\varphi\cos\omega)^2}$ $\omega = \cos^{-1}(-\tan\varphi\tan\delta)$

Table 4.
Solar radiation diffuse factor, R_d, expressed in terms of the tilt angle, β. Modified from [61].

intensity of diffuse sky radiation to be uniform over the skydome. As a consequence, the diffuse radiation incident on a tilted surface is a function of the fraction of the skydome it sees. The anisotropic models on the other hand assume the anisotropy of the diffuse sky radiation in the circumsolar region (portion of sky near the solar disk) plus and isotropically distributed diffuse component from the rest of the skydome. Some of these models are summarised in **Table 4**.

4.4 Optimization of Tilt angle techniques

For PV modules to furnish maximum output power, there is a need to optimize the tilt angle. We present here (**Table 5**) a non-exhaustive summary of some optimal tilt angle equations while the details are obtainable from the indicated references. Taking into consideration the functional relationship of the solar declination, δ, with the day of the year through (Eq. (7)), we also include the optimum tilt angle data (monthly, seasonally and yearly) as applicable alongside the models (**Tables 4** and **5**).

5. Conclusion

In this chapter, we have presented the different models of SR geared towards photovoltaic applications. Solar radiation models can be distinguished based on the type of measurement of input data used. Based on this we have models that use ground measured data and models that use satellite measured data. These models can be further sub-classified as either broadband or spectral models according as they are based on the earth's radiation balance or on results generated by the

Reference	Location	Optimum tilt angle equation	Optimum tilt angle Interval	Data
[74]	Romania	$\beta_{opt} = \varphi - \delta$	/	/
[75]	Ghana	$\beta_{opt} = \varphi - 17^0$	Monthly	40^0 (Jan.); 28.7^0 (Feb.); 10.6^0 (march)
			Yearly	26.8^0
[76]	Brisbane, Australia	NA	Yearly	26^0 N
[77]	Abu Dhabi, UAE	NA	Monthly	June -9^0, December 52^0
[78]	8 provinces of Turkey	$\beta_{opt} = 34.783 - 1.4317\delta - 0.00816\delta^2 + 0.0002 \delta^3$	Monthly	$0°$ (June) and $65°$ (December)
			Seasonally	$21.17°$ (spring), $5.67°$ (summer), $46.48°$ (autumn), $57.29°$ (winter)
[79]	Ontario, Canada	$\beta_{opt} = \varphi$	Yearly	$\beta_{opt} = \varphi$
	Ottawa, Canada	$\beta_{opt} = \varphi - (7^0 \to 12^0)$	Yearly	$\beta_{opt} = 36^0 \to 38^0; \varphi = 45^0$
	Toronto, Canada	$\beta_{opt} = \varphi - (7^0 \to 12^0)$	Yearly	$\beta_{opt} = 32^0 \to 35^0; \varphi = 44^0$
[80]	United States	$\beta_{opt} = \varphi - (1^0 \to 10^0)$	Yearly	Optimum orientation (Tilt/Azimuth) Orlando, FL, 29.1°/7.6°E; Dallas, TX, 30.5°/5.9°E; Phoenix, AZ, 33.4°/0.3°W; Los Angeles, CA, 32.4°/3.8°W; St. Louis, MO, 34.8°/1.0°W
[81]	Zahedan, Iran	$\beta_{opt} = 09.17\varphi \pm 0.321^0$	Bi-Annual	$5^0, 50^0$
[82]	Madinah, Saudi Arabia	$\beta_{opt} = \varphi$	Monthly	40^0 (January); 10^0 (June); 40^0 (December)
			Seasonally	37^0 (Winter); 12^0 (Summer); 17^0 (Spring); 28^0 (Autumn)
			Yearly	23.5^0, Equal to latitude, $\beta_{opt} = \varphi$
[83]	New Delhi, India	Summer; $\beta_{opt} = \varphi - 60^0$ Winter; $\beta_{opt} = \varphi + 90^0$	Seasonally	Summer; $13^0 (\varphi - 60^0)$ Winter; $47.5^0 (\varphi + 90^0)$

Table 5.
Solar radiation diffuse factor R_d, with respect to tilt angle β. Modified from [61].

solution of the radiative transfer equation. The baseline objective of the SR models presented is to predict the three components of GSR (the beam, the diffuse and the reflected components) incident on some PV collector surface at the ground level. Consequently, in the models presented, careful attention was given to show how we could quantify these three components. Results for some geographical locations have been given to show the correlation of equations for the models based on the diffuse ratio- clearness index regressions as well as for those based on sunshine fraction.

The broadband models were prioritized in the presentation given the ubiquitous occurrence of the input data for such models. Space restrictions conditioned the presentation of the models to be summarised to the strict minimum so our readers are encouraged to consult the cited literature in addition. A comparison of the models has been presented to highlight the input and the output parameters. In addition, a flow chart to show the interrelatedness of the models has been presented.

The approaches for getting the optimal tilt and azimuthal angles of the PV panel are summarized. Based on literature sources, the optimal tilt angles for some global geographical locations have been presented.

The statistical procedures to ascertain the validity of the regression analyses are summarily treated and applied in some cases where results have been presented.

We expect this chapter to be a valuable tool for scientist and engineers specialized in solar PV research and applications.

Acknowledgements

The authors wish to acknowledge the Cameroon Ministry of Higher Education for financing this research through the research allowance paid to all its staff of Higher Education in Cameroon.

Conflict of interest

No conflict of interest.

Author details

David Afungchui[1*], Joseph Ebobenow[2], Ali Helali[3]
and Nkongho Ayuketang Arreyndip[4]

1 Department of Physics, Faculty of Sciences, The University of Bamenda, Bambili, NWR, Cameroon

2 Department of Physics, Faculty of Sciences, University of Buea, SWR, Cameroon

3 National Engineering School of Sousse, Higher Institute of Transport and Logistics, Mechanical Laboratory, University of Sousse, Riadh City, Sousse, Tunisia

4 Potsdam Institute for Climate Impact Research (PIK), Potsdam, Germany

*Address all correspondence to: afungchui.david@ubuea.cm

IntechOpen

References

[1] M. Bortolini, M. Gamberi, A. Graziani, R. Manzini, and C. Mora, "Multi-location model for the estimation of the horizontal daily diffuse fraction of solar radiation in Europe," *Energy Conversion and Management*, vol. 67, pp. 208-216, 2013.

[2] J. K. Page, "The estimation of monthly mean values of daily total shortwave radiation on vertical and inclined surfaces from sunshine records for latitudes 40° N-40° S," *Proc. UN Conf. New Sources Energy*, Conference Proceeding vol. 4, no. null, p. 378, 1961.

[3] A. Angstrom, "Solar and terrestrial radiation. Report to the international commission for solar research on actinometric investigations of solar and atmospheric radiation," *Quarterly Journal of the Royal Meteorological Society*, vol. 50, no. 210, pp. 121-126, 1924.

[4] K. Mehmet, "Estimation of global solar radiation on horizontal surface in Erzincan, Turkey," *International Journal of Physical Sciences*, vol. 7, no. 33, pp. 5273-5280, 2012.

[5] A. David, E. Joseph, N. R. Ngwa, and N. A. Arreyndip, "Global Solar Radiation of some Regions of Cameroon using the Linear Angstrom Model and Non-linear Polynomial Relations: Part 2, Sun-path Diagrams, Energy Potential Predictions and Statistical Validation," *International Journal of Renewable Energy Research (IJRER)*, vol. 8, no. 1, pp. 649-660, 2018.

[6] C. Ma and M. Iqbal, "Statistical comparison of solar radiation correlations Monthly average global and diffuse radiation on horizontal surfaces," *Solar Energy*, vol. 33, no. 2, pp. 143-148, 1984.

[7] M. Tiris, C. Tiris, and I. Türe, "Correlations of monthly-average daily global, diffuse and beam radiations with hours of bright sunshine in Gebze, Turkey," *Energy Conversion and Management*, vol. 37, no. 9, pp. 1417-1421, 1996.

[8] H. Bulut and O. Büyükalaca, "Simple model for the generation of daily global solar-radiation data in Turkey," *Applied Energy*, vol. 84, no. 5, pp. 477-491, 2007.

[9] R. Stone, "Improved statistical procedure for the evaluation of solar radiation estimation models," *Solar energy*, vol. 51, no. 4, pp. 289-291, 1993.

[10] A. David and N. R. Ngwa, "Global Solar Radiation of some regions of Cameroon using the linear Angstrom and non-linear polynomial relations (Part I) model development," *International Journal of Renewable Energy Research (IJRER)*, vol. 3, no. 4, pp. 984-992, 2013.

[11] I. T. Toğrul and H. Toğrul, "Global solar radiation over Turkey: comparison of predicted and measured data," *Renewable Energy*, vol. 25, no. 1, pp. 55-67, 2002.

[12] M. Iqbal, "Correlation of average diffuse and beam radiation with hours of bright sunshine," *Solar Energy*, vol. 23, no. 2, pp. 169-173, 1979.

[13] A. Kilic and A. Ozturk, "Solar Energy," ed: Kipas Publishing Inc, 1983.

[14] M. Iqbal, *An Introduction To Solar Radiation*. Elsevier Science, 2012.

[15] T. Lealea and R. Tchinda, "Estimation of diffuse solar radiation in the South of Cameroon," *Journal of Energy Technologies and Policy*, vol. 3, no. 6, pp. 32-42, 2013.

[16] P. Cooper, "The absorption of radiation in solar stills," *Solar energy*, vol. 12, no. 3, pp. 333-346, 1969.

[17] J. A. Duffie and W. A. Beckman, *Solar Engineering of Thermal Processes* (null). 1991, p. null.

[18] S. E. Tuller, "Relationship between diffuse, total, and extraterrestrial solar radiation," *Sol. Energy;(United States)*, vol. 18, no. 3, 1976.

[19] B. Y. Liu and R. C. Jordan, "The interrelationship and characteristic distribution of direct, diffuse and total solar radiation," *Solar energy*, vol. 4, no. 3, pp. 1-19, 1960.

[20] D. Erbs, S. Klein, and J. Duffie, "Estimation of the diffuse radiation fraction for hourly, daily and monthly-average global radiation," *Solar energy*, vol. 28, no. 4, pp. 293-302, 1982.

[21] G. Stanhill, "Diffuse sky and cloud radiation in Israel," *Solar Energy*, vol. 10, no. 2, pp. 96-101, 1966.

[22] C. N. Rao, W. A. Bradley, and T. Y. Lee, "The diffuse component of the daily global solar irradiation at Corvallis, Oregon (USA)," *Solar Energy*, vol. 32, no. 5, pp. 637-641, 1984.

[23] B. Bartoli, V. Cuomo, U. Amato, G. Barone, and P. Mattarelli, "Diffuse and beam components of daily global radiation in Genova and Macerata," *Solar Energy*, vol. 28, no. 4, pp. 307-311, 1982.

[24] B. Liu and R. Jordan, "Daily insolation on surfaces tilted towards equator," *ASHRAE J.;(United States)*, vol. 10, 1961.

[25] M. Iqbal, "Prediction of hourly diffuse solar radiation from measured hourly global radiation on a horizontal surface," *Solar energy*, vol. 24, no. 5, pp. 491-503, 1980.

[26] E. L. Maxwell, "A quasi-physical model for converting hourly global horizontal to direct normal insolation," *qpmc*, 1987.

[27] R. Perez, P. Ineichen, R. Seals, and A. Zelenka, "Making full use of the clearness index for parameterizing hourly insolation conditions," *Solar Energy*, vol. 45, no. 2, pp. 111-114, 1990.

[28] A. S. o. H. R. a. A. c. Engineers and (ASHRAE), "Handbook of Fundamentals," ed, 1972, pp. 385-443.

[29] N. Nijegorodov, "Improved ASHRAE model to predict hourly and daily solar radiation components in Botswana, Namibia, and Zimbabwe," *Renewable energy*, vol. 9, no. 1-4, pp. 1270-1273, 1996.

[30] A. Ianetz, V. Lyubansky, E. Evseev, and A. Kudish, "Regression equations for determining the daily diffuse radiation as a function of daily beam radiation on a horizontal surface in the semi-arid Negev region of Israel," *Theoretical and applied climatology*, vol. 69, no. 3-4, pp. 213-220, 2001.

[31] M. S. Gul, T. Muneer, and H. D. Kambezidis, "Models for obtaining solar radiation from other meteorological data," *Solar Energy*, vol. 64, no. 1-3, pp. 99-108, 1998.

[32] M. Bashahu, "Statistical comparison of models for estimating the monthly average daily diffuse radiation at a subtropical African site," *Solar Energy*, vol. 75, no. 1, pp. 43-51, 2003.

[33] M. El-Metwally, "Simple new methods to estimate global solar radiation based on meteorological data in Egypt," *Atmospheric Research*, vol. 69, no. 3-4, pp. 217-239, 2004.

[34] C. Gueymard, "An atmospheric transmittance model for the calculation of the clear sky beam, diffuse and global photosynthetically active radiation," *Agricultural and Forest Meteorology*, vol. 45, no. 3, pp. 215-229, 1989/03/01/ 1989.

[35] M. Gul and T. Muneer, "Solar diffuse irradiance: estimation using air

mass and precipitable water data," *Building Services Engineering Research and Technology,* vol. 19, no. 2, pp. 79-85, 1998.

[36] C. A. Gueymard, "Direct solar transmittance and irradiance predictions with broadband models. Part I: detailed theoretical performance assessment," *Solar Energy,* vol. 74, no. 5, pp. 355-379, 2003.

[37] R. E. Bird, "A simple, solar spectral model for direct-normal and diffuse horizontal irradiance," *Solar energy,* vol. 32, no. 4, pp. 461-471, 1984.

[38] R. E. Bird and C. Riordan, "Simple solar spectral model for direct and diffuse irradiance on horizontal and tilted planes at the earth's surface for cloudless atmospheres," *Journal of Applied Meteorology and Climatology,* vol. 25, no. 1, pp. 87-97, 1986.

[39] C. A. Gueymard, "Parameterized transmittance model for direct beam and circumsolar spectral irradiance," *Solar Energy,* vol. 71, no. 5, pp. 325-346, 2001.

[40] E. Shettle and R. Fenn, "Models for the aerosols of the lower atmosphere and the effects of humidity variations on their optical properties. Air Force Geophysical Laboratory Tech. Rep," *Environmental Research Papers,* vol. 676, p. 94, 1979.

[41] C. A. Gueymard and H. D. Kambezidis, "Solar Spectral Radiation," in *Solar Radiation and Daylight Models,* T. Muneer, Ed. First 1997 ed.: Oxford: Elsevier., 2004.

[42] J. E. Hay, "Satellite based estimates of solar irradiance at the earth's surface —I. Modelling approaches," *Renewable Energy,* vol. 3, no. 4-5, pp. 381-393, 1993.

[43] G. L. Powell, A. J. Brazel, and M. J. Pasqualetti, "New approach to estimating solar radiation from satellite imagery," *The Professional Geographer,* vol. 36, no. 2, pp. 227-233, 1984.

[44] S. Fritz, P. K. Rao, and M. Weinstein, "Satellite measurements of reflected solar energy and the energy received at the ground," *Journal of the Atmospheric Sciences,* vol. 21, no. 2, pp. 141-151, 1964.

[45] V. Ramanathan, "Scientific use of surface radiation budget data for climate studies," in "Surface radiation budget for climate application," 1986, vol. 1169.

[46] K. J. Hanson, "Studies of cloud and satellite parameterization of solar irradiance at the earth's surface," National Oceanic and Atmospheric Administration, Miami, Fla.(USA). Atlantic ... 1971.

[47] J. Ellis and T. Vonder Haar, "Application of meteorological satellite visible channel radiances for determining solar radiation reaching the ground," presented at the Conference on Aerospace and Aeronautical Meteorology, 7th, and Symposium on Remote Sensing from Satellites Melbourne, Fla, November 16-19, 1976, 1977.

[48] J. E. Hay and K. J. Hanson, "A satellite-based methodology for determining solar irradiance at the ocean surface during GATE," *Bull. Amer. Meteor. Sci.,* vol. 59, p. 1549, 1978.

[49] C. Sorapipatana and R. Exell, "An operational system for mapping global solar radiation from GMS satellite data," *ASEAN Journal on Science and Technology for Development,* vol. 5, no. 2, pp. 79-100, 1988.

[50] M. Nunez, "Use of satellite data in regional mapping of solar radiation," in *Szokology, SV, Solar World Congress,* 1983, vol. 4.

[51] M. Nunez, T. Hart, and J. Kalma, "Estimating solar radiation in a tropical

environment using satellite data," *Journal of climatology,* vol. 4, no. 6, pp. 573-585, 1984.

[52] D. Cano, J.-M. Monget, M. Albuisson, H. Guillard, N. Regas, and L. Wald, "A method for the determination of the global solar radiation from meteorological satellite data," *Solar energy,* vol. 37, no. 1, pp. 31-39, 1986.

[53] C. Gautier, G. Diak, and S. Masse, "A simple physical model to estimate incident solar radiation at the surface from GOES satellite data," *Journal of Applied Meteorology and Climatology,* vol. 19, no. 8, pp. 1005-1012, 1980.

[54] G. R. Diak and C. Gautier, "Improvements to a simple physical model for estimating insolation from GOES data," *Journal of Climate and Applied Meteorology,* vol. 22, no. 3, pp. 505-508, 1983.

[55] C. Gautier and R. Frouin, "Satellite-derived ocean surface radiation fluxes," *Advances in Remote Sensing Retrieval Methods,* pp. 311-329, 1985.

[56] R. Pinker and I. Laszlo, "A modified insolation model for satellite observations," *J. Appl. Meteorol,* 1990.

[57] P. Halpern, "Ground level solar energy estimates using geostationary operational environmental satellite measurements and realistic model atmospheres," *Remote sensing of environment,* vol. 15, no. 1, pp. 47-61, 1984.

[58] J. Dave and N. Braslau, "Effect of cloudiness on the transfer of solar energy through realistic model atmospheres," *Journal of Applied Meteorology and Climatology,* vol. 14, no. 3, pp. 388-395, 1975.

[59] W. Möser and E. Raschke, "Incident solar radiation over Europe estimated from METEOSAT data," *Journal of Applied Meteorology and Climatology,* vol. 23, no. 1, pp. 166-170, 1984.

[60] S. Munawwar, "Modelling hourly and daily diffuse solar radiation using world-wide database," Napier University, 2006.

[61] A. Hafez, A. Soliman, K. El-Metwally, and I. Ismail, "Tilt and azimuth angles in solar energy applications–A review," *Renewable and Sustainable Energy Reviews,* vol. 77, pp. 147-168, 2017.

[62] G. A. Kamali, I. Moradi, and A. Khalili, "Estimating solar radiation on tilted surfaces with various orientations: a study case in Karaj (Iran)," *Theoretical and applied climatology,* vol. 84, no. 4, pp. 235-241, 2006.

[63] C. Toledo, A. M. Gracia Amillo, G. Bardizza, J. Abad, and A. Urbina, "Evaluation of solar radiation transposition models for passive energy management and building integrated photovoltaics," *Energies,* vol. 13, no. 3, p. 702, 2020.

[64] M. Hartner, A. Ortner, A. Hiesl, and R. Haas, "East to west–The optimal tilt angle and orientation of photovoltaic panels from an electricity system perspective," *Applied Energy,* vol. 160, pp. 94-107, 2015.

[65] V. Badescu, "A new kind of cloudy sky model to compute instantaneous values of diffuse and global solar irradiance," *Theoretical and Applied Climatology,* vol. 72, no. 1-2, pp. 127-136, 2002.

[66] P. Andersen, "Comments on``Calculations of monthly average insolation on tilted surfaces''by SA Klein," *SoEn,* vol. 25, no. 3, p. 287, 1980.

[67] Y. Tian, R. Davies-Colley, P. Gong, and B. Thorrold, "Estimating solar radiation on slopes of arbitrary aspect," *Agricultural and Forest Meteorology,* vol. 109, no. 1, pp. 67-74, 2001.

[68] P. S. Koronakis, "On the choice of the angle of tilt for south facing solar

collectors in the Athens basin area," *Solar Energy,* vol. 36, no. 3, pp. 217-225, 1986.

[69] D. Reindl, W. Beckman, and J. Duffie, "Evaluation of hourly tilted surface radiation models," *Solar energy,* vol. 45, no. 1, pp. 9-17, 1990.

[70] A. Skartveit and J. A. Olseth, "Modelling slope irradiance at high latitudes," *Solar energy,* vol. 36, no. 4, pp. 333-344, 1986.

[71] M. Steven and M. H. Unsworth, "The angular distribution and interception of diffuse solar radiation below overcast skies," *Quarterly Journal of the Royal Meteorological Society,* vol. 106, no. 447, pp. 57-61, 1980.

[72] J. E. Hay, "Calculation of monthly mean solar radiation for horizontal and inclined surfaces," *Solar energy,* vol. 23, no. 4, pp. 301-307, 1979.

[73] T. M. Klucher, "Evaluation of models to predict insolation on tilted surfaces," *Solar energy,* vol. 23, no. 2, pp. 111-114, 1979.

[74] C. Stanciu and D. Stanciu, "Optimum tilt angle for flat plate collectors all over the World–A declination dependence formula and comparisons of three solar radiation models," *Energy Conversion and Management,* vol. 81, pp. 133-143, 2014.

[75] F. A. Uba and E. A. Sarsah, "Optimization of tilt angle for solar collectors in WA, Ghana," *Pelagia Research Library, Advances in Applied Science Research,* vol. 4, no. 4, pp. 108-114, 2013.

[76] R. Yan, T. K. Saha, P. Meredith, and S. Goodwin, "Analysis of yearlong performance of differently tilted photovoltaic systems in Brisbane, Australia," *Energy conversion and management,* vol. 74, pp. 102-108, 2013.

[77] F. Jafarkazemi and S. A. Saadabadi, "Optimum tilt angle and orientation of solar surfaces in Abu Dhabi, UAE," *Renewable energy,* vol. 56, pp. 44-49, 2013.

[78] K. Bakirci, "General models for optimum tilt angles of solar panels: Turkey case study," *Renewable and Sustainable Energy Reviews,* vol. 16, no. 8, pp. 6149-6159, 2012.

[79] I. H. Rowlands, B. P. Kemery, and I. Beausoleil-Morrison, "Optimal solar-PV tilt angle and azimuth: An Ontario (Canada) case-study," *Energy Policy,* vol. 39, no. 3, pp. 1397-1409, 2011.

[80] M. Lave and J. Kleissl, "Optimum fixed orientations and benefits of tracking for capturing solar radiation in the continental United States," *Renewable Energy,* vol. 36, no. 3, pp. 1145-1152, 2011.

[81] H. Moghadam, F. F. Tabrizi, and A. Z. Sharak, "Optimization of solar flat collector inclination," *Desalination,* vol. 265, no. 1-3, pp. 107-111, 2011.

[82] M. Benghanem, "Optimization of tilt angle for solar panel: Case study for Madinah, Saudi Arabia," *Applied Energy,* vol. 88, no. 4, pp. 1427-1433, 2011.

[83] M. Jamil Ahmad and G. N Tiwari, "Optimization of tilt angle for solar collector to receive maximum radiation," *The open renewable energy journal,* vol. 2, no. 1, 2009.

Forecasting and Characterization of Solar Radiation

Chapter 4

Forecasting and Modelling of Solar Radiation for Photovoltaic (PV) Systems

Ines Sansa and Najiba Mrabet Bellaaj

Abstract

Solar radiation is characterized by its fluctuation because it depends to different factors such as the day hour, the speed wind, the cloud cover and some other weather conditions. Certainly, this fluctuation can affect the PV power production and then its integration on the electrical micro grid. An accurate forecasting of solar radiation is so important to avoid these problems. In this chapter, the solar radiation is treated as time series and it is predicted using the Auto Regressive and Moving Average (ARMA) model. Based on the solar radiation forecasting results, the photovoltaic (PV) power is then forecasted. The choice of ARMA model has been carried out in order to exploit its own strength. This model is characterized by its flexibility and its ability to extract the useful statistical properties, for time series predictions, it is among the most used models. In this work, ARMA model is used to forecast the solar radiation one year in advance considering the weekly radiation averages. Simulation results have proven the effectiveness of ARMA model to forecast the small solar radiation fluctuations.

Keywords: solar radiation, PV power, forecasting, ARMA, fluctuation

1. Introduction

Solar energy is a renewable energy source, clean and inexhaustible. It is based on the photovoltaic effect to convert solar energy into electricity through solar cells. PV panels was mainly installing in isolated areas to provide them the electricity but in the last few years a considerable amount of electricity has been generated from solar energy in different countries in the world. In 2019, the global installed solar energy capacity has reached 586.42 GW [1]. This significant growth will may be continuing in the future due to its several technological, environmental and economic benefits.

Like some other renewable energies, solar energy is intermittent. Its production is so related to the solar radiation received on the earth. Therefore, it is possible to forecast solar energy from a relevant forecasting of solar radiation. Different techniques have been developed in the literature to forecast solar radiation. Most of them treat it as time series. These techniques are based on the historical solar radiation data, they treated and followed the solar radiation evolution on the past. Based on the historical data, a model is created to characterize the solar radiation behavior in the past. Therefore, the forecasting of solar radiation on a given time interval is based on this created model. The aim of this chapter is the forecasting of solar radiation using ARMA model. Based on these results and taking into account some other parameters, the PV power is then modeled. A general overview of solar radiation and its different propagation forms is presented in the first part of this chapter. Then, a brief literature review on solar radiation forecasting techniques will be the subject of the next part. After that, the ARMA model will be used to forecast the annual solar radiation corresponds to an industrial company by considering the weekly radiation averages. PV power is modeled in the following section and based on the forecasting solar radiation results, it is presented for different PV panels number. The end section concludes and summarizes this chapter.

2. General presentation of solar radiation

The sun is a vital element, necessary for photosynthesis, important for plants and fundamental for the thermal balance of different component of the crop. 75% of its composition is Hydrogen and the rest is Helium [2]. The sun is the primary source of electromagnetic radiation in Earth. It emits energy in the form of electromagnetic waves called solar radiation which mainly composed of visible light, ultra violet and infrared radiation. Visible light is the part of electromagnetic spectrum visible with the naked eye, its wavelength is depended to the individual. The ultra violet radiation is characterized by a wavelength greater than 800 nm, it is also called black light. This type of radiation is not visible with the naked eye. The infrared is a radiation with a wavelength less than 400 nm. It is greater than that of visible light but shorter than that of micro wave. When the solar radiation passes through the atmosphere, it is reduced due to its molecular scattering and its absorption by gas molecules. Ultra violet and infrared radiations are the two most absorbed. The amount of energy received on earth is depended to the atmosphere thickness and to some other factors such as the seasonal and cloud variations.

2.1 Propagation of solar radiation in the atmosphere

By propagating in the atmosphere, solar radiation can be diffused, absorbed or reflected,

- Reflected radiation: the radiation is reflected by the earth's surface and the soil reflects the radiation in a diffuse and anisotropic manner.

- Diffused radiation: the radiation is diffused in all direction, this phenomenon is occurred in a medium containing fine molecules and it strongly depends to these molecules size.

- Absorbed radiation: the radiation is absorbed by gas molecules that it encounters in atmosphere, this absorption is mainly due to water vapor, carbon dioxide and ozone.

These different interaction of solar radiation with atmosphere are recapitulated in the **Figure 1** [3].

2.2 Modeling of solar radiation

Several theories are developed in the literature to model solar radiation [4–6]. Therefore, at a specific moment and in a given location, the solar radiation cannot be modeled without requiring some factors such as the sky nature and the sun

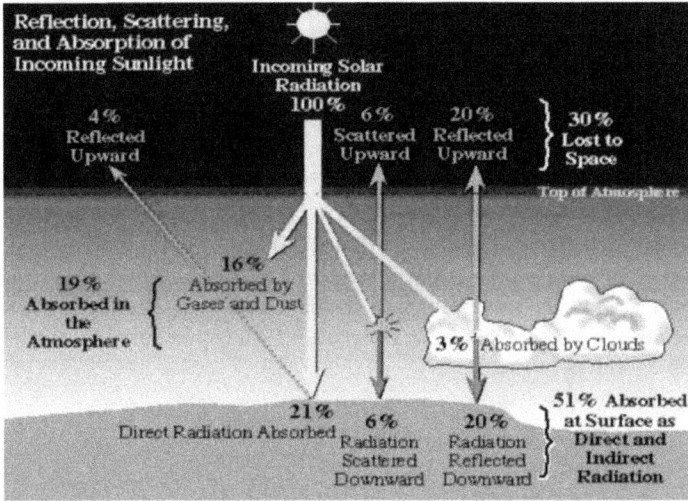

Figure 1.
Interaction between solar radiation and the atmosphere [3].

position. As mentioned previously, solar radiation has three different components, reflected, diffused and absorbed. All these components are modeled by the global or total solar radiation as presented in the Eq. (1).

$$R_{tot} = R_{dir} + R_{dif} + R_{ref} \tag{1}$$

With R_{tot} represents the total solar radiation, R_{dir}, R_{dif} and R_{ref} are respectively the directed, diffused and reflected solar radiation. Each of these radiations is sensitive to certain parameters and are calculated as presented in the following equations,

$$R_{dir} = Sh.Rout.\tau^M. \cos(i) \tag{2}$$

$$R_{diff} = Rout.(0.271 - 0.294.\tau^M).\sin(\alpha) \tag{3}$$

$$R_{ref} = r.Sc.(0.271 + 0.706.\tau^M).\sin(\alpha).\sin^2\left(\frac{x}{2}\right) \tag{4}$$

With Sh is a binary umbrage value, it is computed for each hour in day. Sh is assigned to 0 when the solar radiation is projected to the neighboring mountain umbrage, else it is assigned to 1. r represents the soil reflectance; it is also called the reflection factor. Sc is the constant solar equal to 1367 W/m². To define the other parameters, a recourse to the geometry between sun and earth as well as to the characteristics of the solar flux are needed. Indeed, the position of the sun in the sky depends to the time and latitude. It is defined by two angles which characterized the altitude and the solar azimuth. The altitude angle α is defined as presented in the below Eq. [7].

$$\sin \alpha = \sin \varphi. \sin(\delta) + \cos(\varphi). \cos(\eta) \tag{5}$$

With φ and η are respectively the latitude for each cell and the solar time. δ represents the solar declinaition, this parameter depends to the year day j and expressed as written in Eq. (6),

$$\delta = 23.45. \sin\left(360.\frac{284 + j}{365}\right) \tag{6}$$

The azimuth angle β is defined as presented in the Eq. (7)

$$\cos \beta = (\sin \delta.\cos\varphi - \cos\delta.\sin\varphi.\cos\eta)/\cos\alpha \tag{7}$$

Rout represents the solar flux, it depends to the solar constant Sc and the year day j, it is written as indicated in the Eq. (8),

$$Rout = Sc.\left(1 + 0.034.\cos\left(\frac{360j}{365}\right)\right) \tag{8}$$

τ^M represents the transmissivity coefficient, it is defined as the fraction of the solar radiation incident on the atmosphere surface that reaches the soil along a vertical trajectory. In the mountain area, a correlation factor linked to the atmospheric pressure p/p0 must be used. The path length is presented by the lettre M and written as shown in the Eq. (9),

$$M = M_0.\frac{P}{P_0} \tag{9}$$

M_0 is calculated following the Eq. (10) and the p/p0 represents the correlation factor of atmospheric pressure, it is calculated as defined in the Eq. (11).

$$M_0 = \sqrt{1229 + (614.\sin\alpha)^2} - (614.\sin\alpha) \tag{10}$$

$$\frac{P}{P_0} = \left(\frac{288 - 0.0065.h}{288}\right)^{5.256} \tag{11}$$

An incidence angle i between the sun ray and the soil surface must be taken into account when the solar radiation is converged to sloping areas. This angle is depended to the sun position and to the topography and it is written as described the below equation,

$$\cos i = \cos\alpha.\sin x.\cos(\beta - \beta_s) + \sin\alpha.\cos x \tag{12}$$

With x and β_s represent respectively the slope and the exposure, they are taken in degrees. It should be noted that the Eq. (1) describes the solar radiation without taking into account the clouds effects. To take them into account, a coefficient Kc must be added. So the expression of solar radiation in the presence of clouds R_{totc} will be written as presented in the Eq. (13).

$$R_{totc} = K_c.R_{tot} \tag{13}$$

The K_c coefficient is depended to the cloudiness N and calculated as described in the Eq. (14).

$$K_c = \left(1 - 0.75.\left(\frac{N}{8}\right)^{3.4}\right) \tag{14}$$

3. Forecasting of solar radiation

3.1 Forecast horizon

Before forecasting, it must specify firstly the horizon forecasting. The choice of this horizon is relative to the problem to be treated. They are four forecasting horizon categories which are the very short term, the short term, the medium term and the long term. Each of these horizons is characterized by a time interval as described in the following paragraph,

- Very short term: the time horizon of this forecasting category does not exceed a few hours, usually it is used for the intra-day market.

- Short term: the time horizon of this category is between 48 hours and 72 hours. This type of forecasting horizon is useful for the daily dispatching electrical power.

- Medium term: the time horizon of this forecasting term is done for more than one week to one month. It intervenes in the planning of the power system. It is also used for the dispatching of the conventional power plants.

- Long term: the time horizon of this type is done from one month to one year. It is useful for long term planning operations such as expansion projects for power generation units.

3.2 Solar radiation forecasting techniques

In the literature, different techniques are proposed to the forecasting of solar radiation [7]. It is possible to classify them into four groups, the naïve models, the conditional probability models, the reference models and the connectionist models. A description of each of these techniques is described in the following sub sections.

3.2.1 Naïve model

They are the smallest techniques for time series forecasting. For a given horizon, the forecasting is based on the last observed variable [8]. The mean, the persistence and the k nearest neighbors are registered under these models.

3.2.1.1 Mean forecasting method

The mean forecasting method consists to substitute the variable to be forecasted by the mean available data assigned to this variable. It is a simple technique to apply but it is so expensive in terms of history [7]. If N corresponds to the number of historical data, the forecasting of a variable x at a given horizon h is described as presented in the Eq. (15).

$$\hat{x}_{t+h} = \frac{1}{N} \sum_{i=1}^{N} x_i \tag{15}$$

3.2.1.2 The persistence

This technique is based on the repetition of a measurement from time t to time t+h [7]. If the considered horizon h is 1, the forecasting of a variable data at time time t+1 is defined as written in the Eq. (16).

$$\hat{x}(t+1) = x(t) \tag{16}$$

This predictor type is often used in time series forecasting because it is so easy to implement and it does not require a large historical data base. On the other hand, it is imprecise and it does not lead to an improvement in time series.

3.2.1.3 The k nearest neighbors

This technique is derived from the artificial intelligence, it consists to find in time series, a set of k data similar to those that to be predicted [9]. The determination of k is done by different algorithm [7]. This technique is, in general, efficient in the time series forecasting, however, it is sensitive to the dimensionality and to the irrelevant variables.

3.2.2 Conditional probability models

We cite as an example for these types of models those of Marcov chains and Bayesian inferences.

3.2.2.1 Marcov chain

This technique is rarely used for the forecasting of solar radiation [10]. It is a stochastic process that has the Markovian property [11]. A future state is modeled by a probabilistic process which depends only to the present states. Following Markov chain, the forecasting of a variable at a given horizon h is defined as presented in the Eq. (17).

$$X_{t+h} = X_t . R_M^h \tag{17}$$

With R_M represents the transition matrix of Markov chain, its dimension depends to several factors such as the number of available data and the precision nature [12].

3.2.2.2 Bayesian inferences

This method is mainly based on the conditional probability; it is rarely used for the forecasting of solar radiation. This method is very difficult to handle and it requires several parameters. The estimation of the probability of a series at a given horizon can be done by Bayes theorem as described in the Eq. (18).

$$p(A/B) = \frac{p(B/A) . p(A)}{p(B)} \tag{18}$$

3.2.3 Connectionist models

The first artificial neuron was created by Warren McCulloch and Walter Pitts in 1943 [13]. The structure of this neuron is imitated from the biological neuron as presented in the **Figure 2** [14]. An artificial neural network (ANN), is an assembly strongly connected of formal neurons. It is characterized by an excellent capacity of learning and generalization as well as a speed of processing. Its ability to learn and generalize makes it a very powerful tools. It has proven, in recent years, its effectiveness in various research fields. ANNs are subdivided into two large families, static and dynamic neural network. The choice of the one or the other of these two networks depends to the application to be processed, the available information and the complexity model [15].

3.2.4 Conditional probability models

These are models from the large family of Auto Regressive and Moving Average (ARMA). ARMA is the combination of two models, the Auto Regressive (AR) and Moving Average (MA). It is characterized by its ability to extract useful statistical properties. Thus, it is among the most widely used models for time series forecasting. Its effectiveness to forecast solar radiation is well proven in certain research work [16]. AR model assumes that each point can be forecasted by the sum of p previous points plus a random error term. The expression of AR model with an order p (AR (p)) is written as presented in the Eq. (19),

$$x(t) = \alpha_1.x(t-1) + \alpha_2.x(t-2) + ... \alpha_p.x(t-p) + \varepsilon_t \qquad (19)$$

with α_i represent the AR coefficients and ε_t is a white noise.

The moving average process assumes that each point is the sum of q previous errors plus its own error. The expression of MA model with an order q (MA(q)) is written as presented in the Eq. (20).

$$x(t) = \beta_1.e(t-1) + \beta_2.e(t-2) + ... \beta_q.e(t-q) \qquad (20)$$

With β_i are the MA coefficients. A combination of these two models forms the ARMA model with order p and q, its expression is described in the Eq. (21).

$$x(t) = \alpha_1.x(t-1) + \alpha_2.x(t-2) + ... \alpha_p.x(t-p) + \beta_1.e(t-1)$$
$$+ \beta_2.e(t-2) ... \beta_q.e(t-q) + \varepsilon_t \qquad (21)$$

The major requirement of ARMA model is that the time series studied must be stationary. A series is considered stationary when its statistical properties such the mean and the variance are constant over time [17]. The distribution of a stationary series at time t is identical to that at time t-1. The unit root is among the stationarity tests. Autos-correlations and partial autos-correlations diagrams can be used also to prove the stationarity of time series.

If the time series is proved stationary, an approach must be followed to define the p and q orders. Box and Jenkins methodology is used to determine them, it contains four steps, identification of the model, estimation of the parameters, the validation of the selected model and finally the use of this model for forecasting.

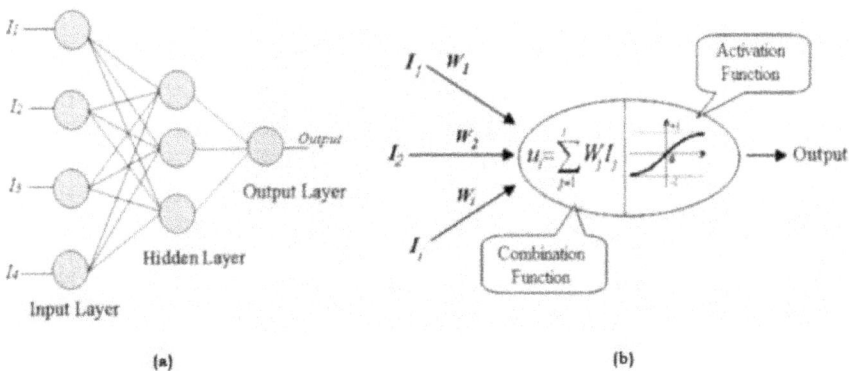

Figure 2.
Schematic diagram of an ANN structure neuron model [14].

- Identification: this is the most important step, it aims to identify the p and q orders. This is done by examining the auto correlation and the partial auto correlation diagrams of the time series.

- Estimation of parameters: the determination of p and q orders does not reflect the validation of this model. It is necessary to estimate the ARMA(p,q) selected. This estimation can be made by the student test.

- Validation model: this validation is carried out by applying two tests on the residues, the Ljung-Box test and homoscedasticity test to ensure that the residuals are white noises.

- The use of model: the selected ARMA model can be used in forecasting. However, in order to ensure the validity of this model, it must be tested on a data base already known. It should find good forecasting performances by comparing the data forecasted by this model and those already known.

3.3 Solar radiation forecasting using ARMA model

The objective of this section is to forecast the solar radiation using ARMA model. The data base solar radiation considered for the forecasting is the set of solar radiation measurements corresponds to an industrial company located in Barcelona north [18]. The time interval of these measurements is five minutes, they are taken every day for a whole year as presented in the **Figure 3**.

To refine the representative curve, just the weekly solar radiation averages are taken into account as presented in the **Figure 4**.

To apply ARMA model, it must study the stationarity of this series. Correlograms corresponding to the auto-correlations and to the partial auto-correlation to this series are presented in the **Figure 5**.

The auto-correlation coefficient of order 1 is close to 1 and the correlogram shows a slow regression which is typical of non-stationary series. Dickey Fuller test is thus applied using EViews software; it proves the weak stationary of this series as shows in the **Figure 6**.

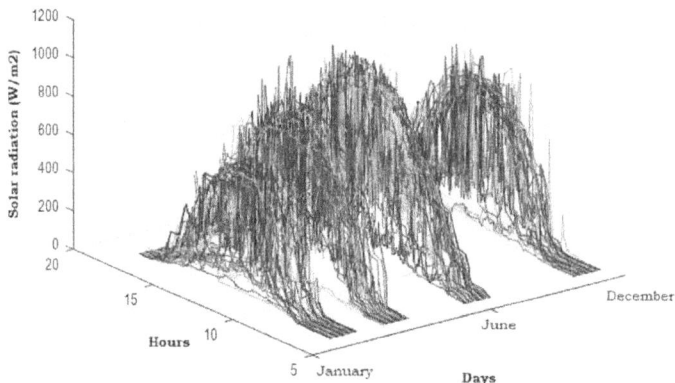

Figure 3.
Annual solar radiation evolution.

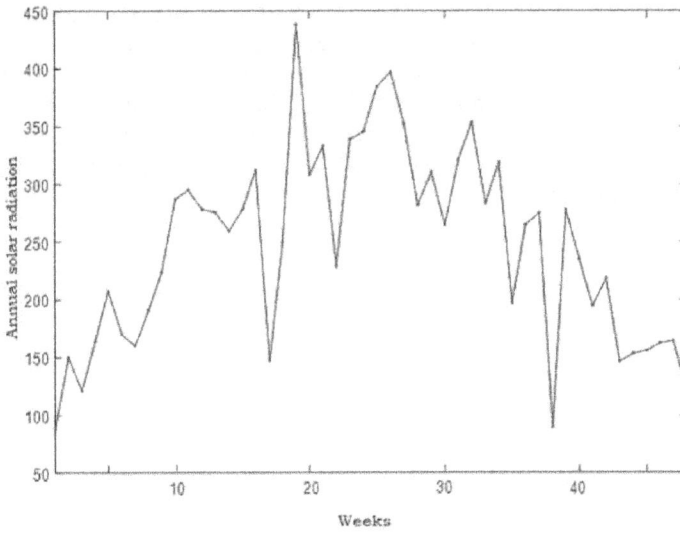

Figure 4.
Weekly solar radiation averages.

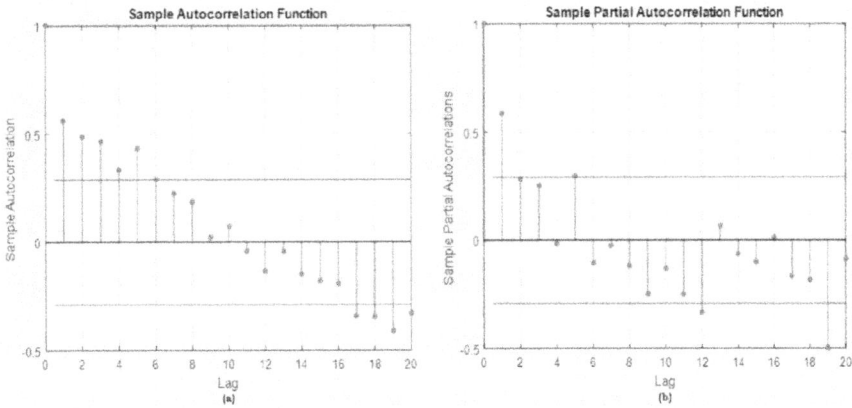

Figure 5.
(a) Auto-correlation and partial auto-correlation (b) correlograms of annual solar radiation.

Figure 6.
Dickey fuller test results for weekly solar radiation series.

The differentiation of this series is necessary in order to make it stationary. The following **Figure** 7 shows the evolution of the differentiated weekly solar radiation.

The Dickey Fuller is thus applied and it proves the stationary of this series as shown in the **Figure 8**. Thereafter, the different Box and Jenkins methodology steps are followed to obtain finally the optimal ARMA model that reproduces the best the behavior of this series. Orders p and q, coefficients α_1, α_2 and β_1 of the ARMA model are recapitulated in the **Table 1**.

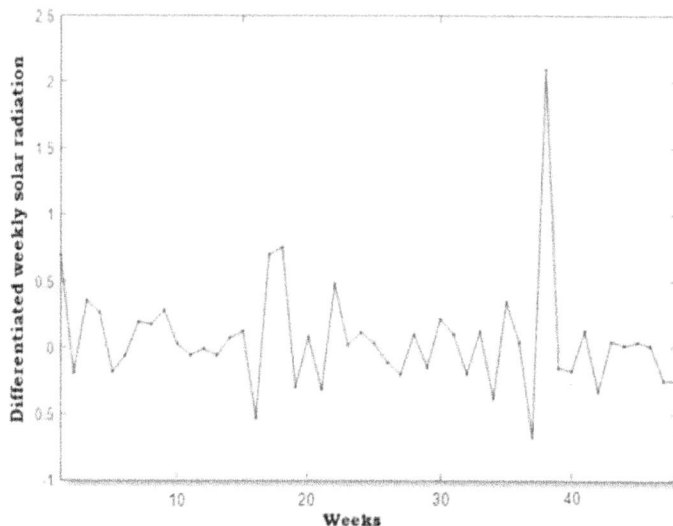

Figure 7.
Differentiated weekly solar radiation evolution.

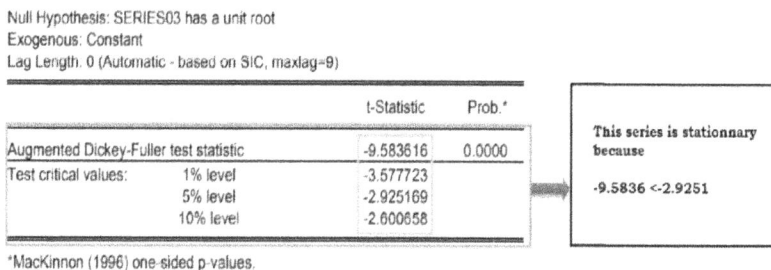

Null Hypothesis: SERIES03 has a unit root
Exogenous: Constant
Lag Length. 0 (Automatic - based on SIC, maxlag=9)

		t-Statistic	Prob.*
Augmented Dickey-Fuller test statistic		-9.583616	0.0000
Test critical values:	1% level	-3.577723	
	5% level	-2.925169	
	10% level	-2.600658	

This series is stationnary because

-9.5836 <-2.9251

*MacKinnon (1996) one-sided p-values.

Figure 8.
Dickey fuller test results for the differentiated weekly solar radiation series.

ARMA (p,q)	
Order	**Coefficients**
p = 2	$\alpha_1 = -1.0342$; $\alpha_2 = -0.4023$
q = 1	$\beta1 = 0.7483$

Table 1.
Orders and coefficients of ARMA (2,1).

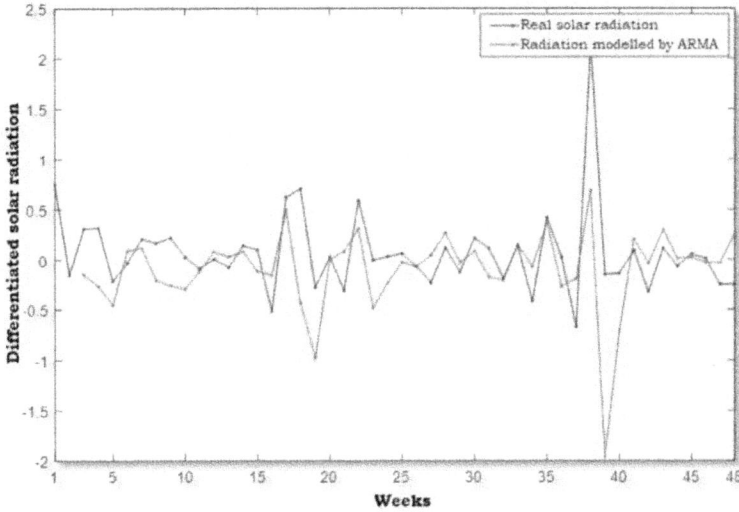

Figure 9.
Solar radiation modeled by ARMA (2,1).

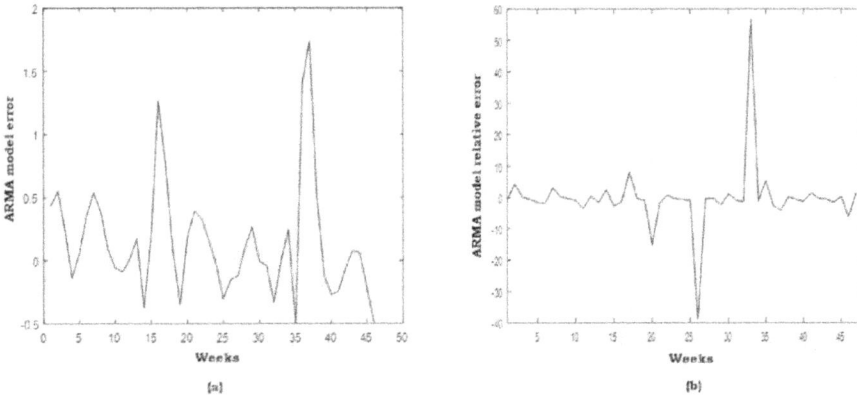

Figure 10.
Error (a) and relative error (b) of solar radiation modeled by ARMA (2,1).

In this paragraph, ARMA (2,1) model is used to forecast the differentiated weekly solar radiation averages. The real solar radiation curve and the forecasted one are presented in the **Figure 9**. It is clear that an approximation is observed between the two curves for certain time intervals, especially when the solar radiation does not present large fluctuations. For other moments time, the forecasted solar radiation curve diverges from the real one. This is particularly observed when the solar radiation presents large fluctuations. To confirm these results, the ARMA model errors are presented in **Figure 10**.

Following the **Figure 10b**, it is clear that the relative error is small, it does not exceed 15%. It is thus observed two peaks, the first one corresponds to the 16th week of the year and the second is in the 36th week. Therefore, when we refer to the real annual solar radiation curve, we observe a sudden fluctuation during these two weeks. Indeed, 16th and 17th weeks correspond respectively to the last week of April month and the first week of May. A considerable decrease of temperature is

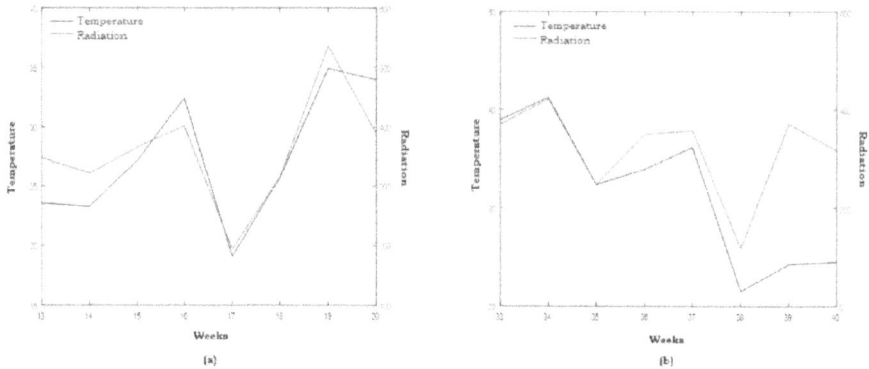

Figure 11.
Influence of temperature on solar radiation (a) April and may weeks (b) September and October weeks.

Errors	Performances
MSE	0.2182
MAE	0.2999
RMSE	0.4671

Table 2.
Errors of solar radiation forecasting using ARMA (2,1).

observed during this period; this is may be the main reason to the sudden decrease of radiation. On the other hand, 37th and 38th weeks correspond to the two first weeks of September month. At the end of this period, it is observed also a sudden decrease of the temperature which affects considerably the solar radiation. Furthermore, as the weekly solar radiation averages are considered for the forecasting, it is obvious to have these large solar radiation variations especially in the switching periods from one season to another one. The influence of temperature on solar radiation evolution for April, May, September and October months are presented in the **Figure 11**.

After forecasting solar radiation or any other parameters, a forecasting error should always be calculated. An error in the forecasting context does not indicate a fault or an anomaly as it is known in several other fields but rather a criterion to evaluate the forecasting performances. In this study, Mean Square Error (MSE), Mean Absolute Error (MAE) and Root Mean Square Error (RMSE) are calculated as written in the Eqs. (22)–(24) [19]. e_i $_{(i=1....n)}$ represents the error measured between the actual value and the forecasted one for sample i and n is the total number of samples. Results are recapitulated in the **Table 2**, the MSE presents the lowest one (0.2182), it is a small value which reflects the performances of ARMA (2,1) model to forecast the solar radiation.

$$\text{MSE} = \frac{1}{n}\sum_{i=1}^{n} e_i^2 \qquad (22)$$

$$\text{RMSE} = \sqrt{\frac{1}{n}\sum_{i=1}^{n} e_i^2} \qquad (23)$$

$$\text{MAE} = \frac{1}{n}\sum_{i=1}^{n} |e_i| \qquad (24)$$

4. Modeling and forecasting of PV power

The forecasting of PV power has a great importance to the best management of grid connected PV systems as well as to the isolated micro grid which include PV system as renewable energy source. Based on the literature, it is possible to forecast the PV power by direct or indirect methods [20]. Direct methods consist to describe models to directly forecast the amount of PV power or forecast the PV power without using other metrological data. In this context, different approaches are suggested which mainly the ANN and the machine learning techniques [20–22]. On the other hand, the indirect methods consist to forecast the PV power based on the forecasting of another meteorological data such as the solar radiation or the temperature [20, 23]. Different physical and statistical approaches are proposed in this field. The choice for the one or the other method is depended to the available data and the forecast horizon term. In physical approaches, the PV power forecasting is based on weather variables predicted by numerical weather prediction (NWP) models and they are more suitable for the long term horizon. The statistical approaches are based on past measured time data series and generally they are appropriate for short term horizon. Moreover, the statistical approaches are simpler than the physical approaches since they require less input data and lower computation [24].

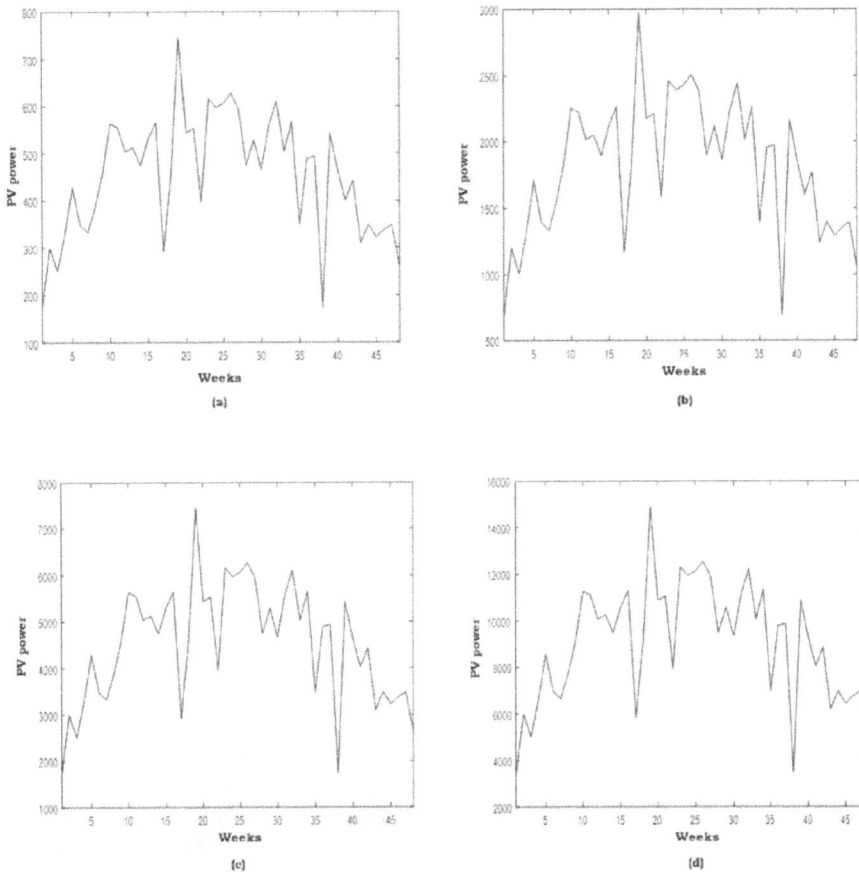

Figure 12.
PV power forecasting for different PV panels number (a) N=5 (b) N=20 (c) N=50 (d) N=100.

In the following section, the PV power will be modeled and forecasted based on the results of solar radiation forecasting, presented in the precedent section. Indeed, the PV power generators are very often operating with a maximum power called Maximum Power Point Tracker (MPPT) [25]. The maximum power P_{PV} delivered by a PV generator composed of N PV panels can be expressed as indicated in the Eq. (25) [26].

$$P_{PV} = \eta_g.N.A.G \tag{25}$$

With A represents the area of a single PV panel, it is expressed in m². G is the solar radiation measured in w/m². η_g is the PV generator efficiency and it is described as written in the Eq. (26) [27].

$$\eta_g = \eta_r.\eta_{pt}.[1 - \beta_t.(T_c - T_r)] \tag{26}$$

η_r represents the reference efficiency of PV generator, it depends to the PV cells materials. η_{pt} is the efficiency of power tracking equipment, it is equal to 1 if the MPPT is perfectly used, β_t is the temperature coefficient, it is expressed in °C. The typical value of this coefficient varies between 0.004 and 0.006, usually, it is taken in the range of 0.005°C [26]. T_c and T_r represent respectively the tcemperature measured in the PV cells and the reference temperature. T_c depends to the ambient temperature T_a and the radiation G as presented in the Eq. (27) [26].

$$T_c = T_a + G.\frac{NOCT - 20}{800} \tag{27}$$

The typical NOCT value for polycrystalline cells is around 45°C. Taking into account Eq. (26) and Eq. (27), the PV power is described as presented in the Eq. (28).

$$P_{PV} = \eta_r.\eta_{pt}.\left[1 - \beta_t.\left(T_a + G.\frac{NOCT - 20}{800} - T_r\right)\right].N.A.G \tag{28}$$

As shown in the Eq. (28), the evolution of PV power depends to several parameters such as the temperature, the solar radiation and the PV panels number. Therefore, it is possible to forecast the PV power from the solar radiation forecasting. So, if the PV cells used is the pollicrystalline and the area of a single PV panel is 2.25m², the evolution of PV power for different PV panels number and based on the solar radiation forecasting results is described as presented in the **Figure 12**.

5. Conclusion

This chapter focuses to model and to forecast the PV power based on the solar radiation forecasting results. Some physical equations are presented firstly to define in general the three different forms of solar radiation. They are explained taking into account some topographical factors and geometric relations.

For solar radiation forecasting, a set of solar radiation measurements corresponds to an industrial company is considered as data base. ARMA model is used to forecast the weekly solar radiation averages. The simulation results obtained are proven the effectiveness of this model to forecast the small variation of solar radiation. On the other hand, it is observed the deterioration of ARMA model with the large solar radiation fluctuations. The forecasting of PV power is carried out based on the obtained solar radiation forecasting results and taking into account some other parameters such as the temperature, the PV cells materials and the PV panels number.

Acknowledgements

This work was supported by Tunisian Ministry of Higher Education and Scientific Research under Grant LSE-ENIT-LR 11ES15.

Author details

Ines Sansa[1*] and Najiba Mrabet Bellaaj[2]

1 Université de Tunis El Manar, Ecole Nationale d'Ingénieurs de Tunis, LR11ES15, Laboratoire des Systèmes Electriques, Tunis, Tunisie

2 Université de Tunis El Manar, Institut Supérieur d'Informatique, Ariana, Tunisie

*Address all correspondence to: sansa.ines@yahoo.com

IntechOpen

References

[1] Statistical Review of World Energy, h ttps://www.bp.com/en/global/corpora te/energy-economics/statistical-revie w-of-world-energy.html

[2] RJ. Aguiar, M. Collares-Pereira, JP. Conde: Simple procedure for generating sequences of daily radiation values using a library of Markov transition matrices, Solar Energy n°40(3), pp 269-279, 1988.

[3] Sultana N. Nahar: Solar irradiation of the earth's atmosphere, Department of Astronomy, the Ohio State University, Columbus, Ohio, USA.

[4] K. Liou: An introduction to atmospheric radiation. Academic Press, 2nd Edition ELSEVIER STORE, http:// store.elsevier.com/An-Introduction-to- Atmospheric Radiation/K_-N_-Liou/ isbn-9780080491677/

[5] F. Kasten: The Linke Turbidity Factor Based on Improved Values of the Integral Rayleigh Optical Thickness, Solar Energy n°56(3), Pp 239–244, 1996.

[6] V. Badescu: Modeling Solar Radiation at the earth's Surface: Recent Advances, Springer, 2008.

[7] C. Voyant: Prédiction de séries temporelles de rayonnement solaire global et de production d'énergie photovoltaïque à partir de réseaux de neurones artificiels, Thèse de doctorat, Spécialité Energétique, Novembre 2011.

[8] L. Ferrara: Méthodes autoprojectives de prévision des séries chronologiques, Modélisation Appliquée Polycopié de Cours, Université Paris Ouest, Février 2011.

[9] M. Sharif, D.H. Burn: Simulating climate change scenarios using an improved k nearest neighbor model, Journal of hydrology, pp. 179-196, 2006.

[10] C. Paoli, C. Voyant, M. Muselli, M. L. Nivet: Multi-horizon irradiation forecasting for Mediterranean locations using time series models, ISES Solar World Congress, 2013

[11] M. Muselli, P. Poggi, G. Notton: A. Louche first order Markov chain model for generating synthetic "typical days" series of global irradiation in order to design PV stand alone systems, Energy conversion and management, Vol 42-6, pp. 675-687, 2001.

[12] C. Piedallu, J.C. Gégout: Multiscale computation of solar radiation for predictive vegetation modelling, Annals of Forest Science 64, pp. 899-909, 2007.

[13] N. Mrabet Bellaaj: Contribution à l'identification et à la commande numériques de la machine asynchrone Algorithmes Génétiques, Réseaux de Neurones et Logique Floue, Thése de doctorat de l'Ecole Nationale d'Ingénieurs de Tunis, 2001.

[14] Utpal Kumar. Dasa, Kok Soon. Teya, Mehdi. Seyedmahmoudiana, Saad. Mekhilefb, Moh Yamani. Idna Idrisc, Willem van. Deventerc, bend. Horanc, Alex. Stojcevski: Forecasting of photovoltaic power generation and model optimization: A review, Renewable and Sustainable Energy Reviews 81, pp. 912-928, 2018.

[15] G. Salah Eddine: Identification des Systèmes non Linéaires par réseaux de neurones, Mastère en Automatique, Université Mohamed Khider-Biskra, Faculté des Sciences et de la technologie, Département: Génie électrique

[16] J. Wu, C.K. Chan: Prediction of hourly solar radiation using a novel hybrid model of ARMA and TDNN, Solar Energy 85, pp. 808-817, 2011.

[17] K. Thibodeau: Application de la méthodologie Box-Jenkins aux séries du ministère de la santé, Maîtrise en Mathématiques et Informatique

Appliquées, Université du Québec, Avril 2011.

[18] Sant Joan les Fonts, Garrotxa, http://www.noel.es/

[19] Ines SANSA: Optimization d'un micro reseau electrique selon la charge d'un site isolé et prediction de la puissance PV, thèse de Doctorat de l'école nationale d'ingenieurs de Tunis.

[20] Abdelhakim. El hendouzi, Abdennaser. Bourouhou: Solar Photovoltaic Power Forecasting, Journal of Electrical and Computer Engineering, 2020.

[21] R. Ahmed, V. Sreeram, Y. Mishra, M. D. Arif: A review and evaluation of the state-of-the-art in PV solar power forecasting: Techniques and optimization, Renewable and Sustainable Energy Reviews, vol. 124, 2020.

[22] Mellit, A. Massi Pavan, E. Ogliari, S. Leva, V. Lughi: Advanced methods for photovoltaic output power forecasting: A review, Applied Sciences, vol. 10, no 2, pp. 487, 2020

[23] J. Antonanzas, N. Osorio, R. Escobar, R. Urraca, F. J. Martinez-de-Pison, F. Antonanzas-Torres: Review of photovoltaic power forecasting, Solar Energy, vol. 136, pp. 78–111, 2016.

[24] Maria. Grazia De Giorg, Paolo Maria. Congedo, Maria. Malvoni: Photovoltaic Power Forecasting Using Statistical Methods: Impact of weather data, IET Science Measurement and Technology, May 2014.

[25] T. Logenthiran, D. Srinivasan, A. M. Khambadkone, T. S.raj: Optimal sizing of an islanded microgrid using evolutionary strategy, IEEE International Conference on Congress on Evolutionary Computation (CEC) Probabilistic Methods Applied to Power Systems (PMAPS), 2010.

[26] S. Missaoui: Prédiction de la production de la puissance PV à l'aide des réseaux de neurones dynamiques. Mastère en Systèmes Electriques, ENIT, Juillet 2012.

[27] D. Abbes, A. Martinez, G. Champenois: Life cycle cost, embodied energy and loss of power supply probability for the optimal design of hybrid power systems, Mathematics and Computers in simulation, 2013.

Chapter 5

Temporal Fluctuations Scaling Analysis: Power Law of Ramp Rate's Variance for PV Power Output

Maina André and Rudy Calif

Abstract

The characterization of irradiance variability needs tools to describe and quantify variability at different time scales in order to optimally integrate PV onto electrical grids. Recently in the literature, a metric called nominal variability defines the intradaily variability by the ramp rate's variance. Here we will concentrate on the quantification of this parameter at different short time scales for tropical measurement sites which particularly exhibit high irradiance variability due to complex microclimatic context. By analogy with Taylor law performed on several complex processes, an analysis of temporal fluctuations scaling properties is proposed. The results showed that the process of intradaily variability obeys Taylor's power law for every short time scales and several insolation conditions. The Taylor power law for simulated PV power output has been verified for very short time scale (30s sampled data) and short time scale (10 min sampled data). The exponent λ presents values between 0.5 and 0.8. Consequently, the results showed a consistency of Taylor power law for simulated PV power output. These results are a statistical perspective in solar energy area and introduce intradaily variability PV power output which are key properties of this characterization, enabling its high penetration.

Keywords: nominal variability, power Taylor law, intradaily variability, temporal fluctuations scaling, PV power output

1. Introduction

Solar energy is an environmental process composed of a stochastic component, source of this intermittent nature and a deterministic component depending on solar geometry and time/location parameters. The stochastic component is complex to define due to significant fluctuations, particularly at intradaily time scales or short time scales. This component is the result of several factors of clouds motion and weather systems and is the main source of limited penetration onto electrical grids of systems exploiting solar energy such as photovoltaic panels (PV systems). Recently in literature, irradiance short-term variability attracted the interest of many studies. Indeed, the variability of irradiance particularly at short time scales is a very complex process that needs tools to characterize it to optimally integrate it onto electrical grids. The dynamic of fluctuations remains a challenging parameter to define. Several works defined this dynamic by metrics. In [1], a scoring method, termed an Intra-Hour Variability Score (IHVS), quantified variability characteristics into a single metric which represents an hour of irradiance. In [2], the one-minute intra-hourly solar variability based upon hourly inputs has defined four metrics characterizing intra-hourly variability, such as the standard deviation of the global irradiance clear sky index, the mean index change from one-time interval to the next, the maximum and the standard deviation of the latter. Other metrics defining intradaily of irradiance are described in the literature such as VI index (variability index) with the daily clear sky index in [3], the daily probability of persistence (POPD) in [4], the nominal variability which is the ramp rate standard deviation calculated from the change in the clear sky index developed in [5], MAD metric which is defined by the median absolute deviation of the change in the clear sky index in [6].

Analysis of variability was also applied to PV power output such as [7] who defined a frequency domain of PV output variability analysis, or [8] describing the frequency of a given fluctuation from PV power output for a certain day by an analytic model and [9] which demonstrated rapid ramps observed in point measurements would be smoothed by large PV plants and the aggregation of multiple PV plants with [5] who completed and strengthened this result. In [10], a quantitative metric called the Daily Aggregate Ramp Rate (DARR) is proposed to quantify, categorize, and compare daily variability from power output, across multiple sites.

In this chapter, we examine a temporal scaling fluctuation modeling namely power Taylor law applied to irradiance intradaily variability and PV power output intradaily variability. The influence of parameters such as increment, data sampling on this modeling is also assessed in order to reinforce the quantification and characterization of this complex process which is ramp rate's variance. This study is a supplementary results to works about intradaily variability quantification but also showed evidence to the universality of power Taylor for environmental complex processes.

2. Data set for the study

2.1 Context of study

In this work, the sites under study are located in tropical islands (Guadeloupe, La Reunion and Hawaii). These exhibit high variability irradiance due to a large diversity of microclimates. This complex process evolves on different time and spatial scales. **Table 1** summarizes the description of sites under study and **Figure 1**

	Reunion		Hawaii	Guadeloupe
	Saint-pierre	Tampon	Kalaeloa Oahu	Pointe-à-pitre (Fouillole)
Data provider	PIMENT	PIMENT	NREL	LARGE
GHI measurement time step	10 min	10 min	3 seconds	1 second
Period of record	2 years	2 years	2 years	2 years
Longitude (°)	55.491	55.506	−158.084	−61.517
Latitude (°)	−21.34	−21.269	21.3120	16.217
Elevation (m)	75	550	11	6

Table 1.
Description of sites under study.

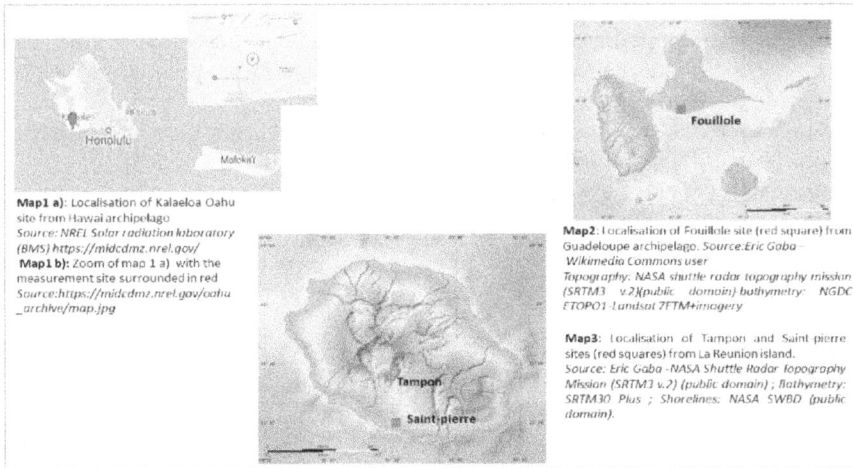

Map1 a): Localisation of Kalaeloa Oahu site from Hawai archipelago
Source: NREL Solar radiation laboratory (BMS) https://midcdmz.nrel.gov/
Map1 b): Zoom of map 1 a) with the measurement site surrounded in red
Source:https://midcdmz.nrel.gov/oahu_archive/map.jpg

Map2: Localisation of Fouillole site (red square) from Guadeloupe archipelago. Source:Eric Gaba - Wikimedia Commons user
Topography: NASA shuttle radar topography mission (SRTM3 v.2)(public domain) bathymetry: NGDC ETOPO1-Landsat 7ETM+imagery

Map3: Localisation of Tampon and Saint-pierre sites (red squares) from La Reunion island. Source: Eric Gaba -NASA Shuttle Radar Topography Mission (SRTM3 v.2) (public domain) ; Bathymetry: SRTM30 Plus ; Shorelines: NASA SWBD (public domain).

Figure 1.
Geographical location of measurement sites under study: Oahu, Fouillole campus, tampon and Saint-Pierre.

presents the geographical location of measurement sites under study. Measurements are available on a basis of two years of data. The study of temporal irradiance fluctuations scaling is therefore analyzed for different locations.

2.1.1 Case of Oahu

Kalaeloa Oahu is located in a tropical zone, at the West of the Hawaii island. This station is affected by clouds formation during summer due to the trade winds effect and are generated by the local topography (located inland with medium orography with an elevation of about 11 m). This dataset is provided on the NREL (National Renewable Energy Laboratory) website. The procedure of data acquisition is described on the website. GHI is measured by using a LICOR LI-200 Pyranometer mounted on an Irradiance Inc. Rotating Shadowband Radiometer (RSR). RSR mounted on the ground and the LI-200 sensor height is approx. The uncorrected value is for testing and troubleshooting purposes only. Voltage is measured across a 100 Ohm precision resistor in parallel to the sensor output.

2.1.2 Case of Fouillole

Fouillole site is located at the campus of the French West Indies University situated in the West of Grande-Terre island in coastal topography and also located in an urban area. This context generates a complex microclimatic context. The clouds are generated by land/sea contrast and the local topography (elevation lower than 10 m, **Table 1**). Data are measured by a pyranometer CM22 from Kipp and Zonen whose response time is less than one second. The precision of pyranometer is $+/-3.0\%$ for the daily sun of GHI. Measurements are provided by LARGE laboratory from Université des Antilles on a 1 second basis data.

2.1.3 Case of Saint-Pierre and Le tampon

Concerning Reunion island, two locations at the West of the island are under our study: Saint-Pierre which is a coastal site, and Le Tampon an inland site. According to [11, 12], these two sites exhibit very different sky conditions. Concerning Le

Tampon, the inland site orographic clouds are mainly generated by the local topography. This site is located in a mountainous orography (elevation about 550 m) in an urban zone. It presents higher variability irradiance than Saint-Pierre site which is in a climate tropical ocean with an urban coastal topography. The irradiance data is measured with a secondary standard pyranometer CMP11 from Kipp and Zonen. The precision of the pyranometer is $+/-3.0\%$ for the daily sun of GHI. Measurements provided by PIMENT are available on a 10 min basis and two years of data.

2.2 Data preprocessing

The profile of GHI that is due to solar geometry is predictable by several models [13–15]. In our study, we will focus on intra-daily variability induced by cloud mass passage that is stochastic in nature [5].

In order to study this variability component, the solar-geometry effects must be first removed. The parameter usually considered in the solar energy area is the clear sky index Kt^* (ratio of measured GHI to theoretical clear sky GHI) defined as Eq. (1).

$$Kt^* = \frac{GHI_m}{GHI_{clear}} \qquad (1)$$

where *GHI* is the Global Horizontal Irradiance, index *m* refers to the measured GHI and index *clear* refers to theoretical clear sky irradiance.

In order to better consider variability for a time scale, we investigate in the temporal increment for a given time scale Δt. The temporal increment of Kt^* corresponding to the selected time scale Δt is noted $\Delta K^*(t, \Delta t)$ and is defined Eq.(2) such as in [5]. A sequence of $\Delta K^*(t, \Delta t)$ for each measurement sites with $\Delta t = 20\,min$ from original Kt^* time series at 10 min, is presented in **Figure 2**. This change is often referred to as the ramp rate [5].

$$\Delta K^*(t, \Delta t) = K^*(t + \Delta t) - K^*(t) \qquad (2)$$

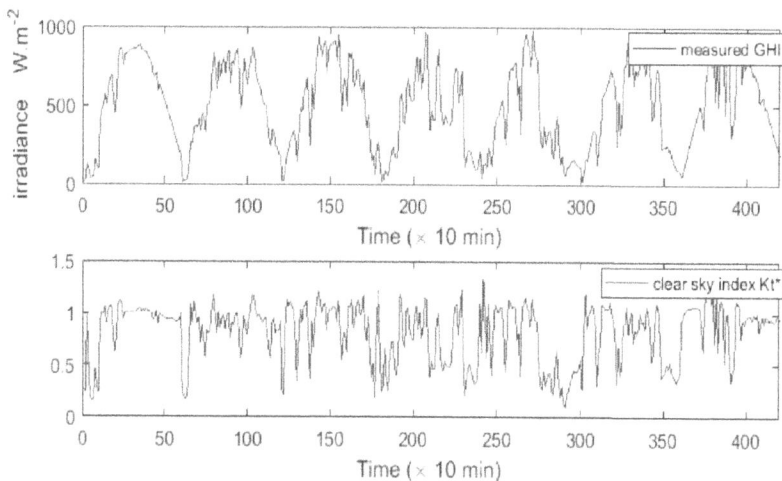

Figure 2.
Signals of $\Delta K^*(t, \Delta t)$ for each measurement sites with $\Delta t = 20\,min$ from original time series of Kt^* at 10 min.

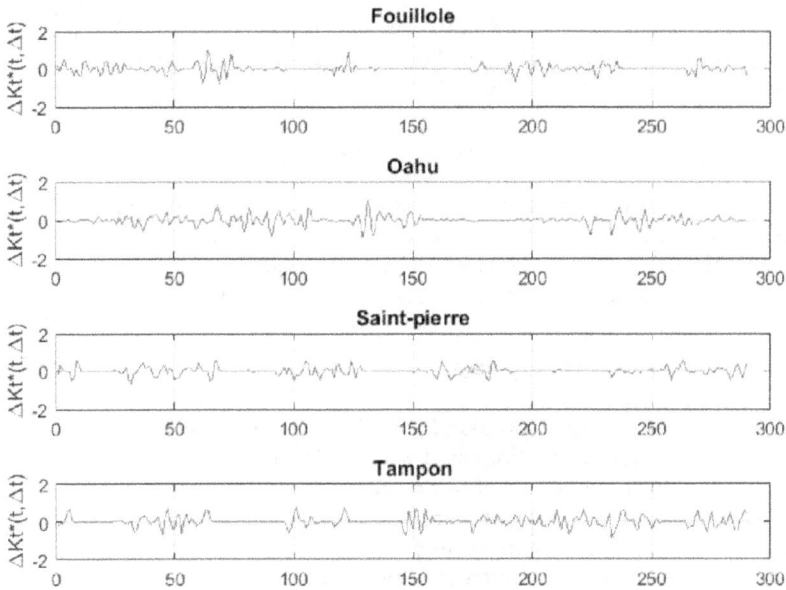

Figure 3.
(a) 1 week GHI data and b) the corresponding Kt* signal obtained for Fouillole site. The latter signal exhibits the high variability of solar flux.

Figure 3 presents Kt^* time series for 1 week sequence obtained from 1 week GHI data at 10 minutes time scales and **Figure 2** presents $\Delta K^*(t, \Delta t)$ time series for five other days sequences.

Recently in the literature [5, 6], a metric is defined to characterize the intradaily variability of the change in the clear sky index over the considered day i.e. the ramp rate's variance, or its square root. This metric is the ramp rate standard deviation called nominal variability defined by this equation:

$$Nominal\ variability = \sigma\left(\Delta K^*(t, \Delta t) = \sqrt{Var[(\Delta K^*(t, \Delta t)]}\right) \qquad (3)$$

This metric can clearly distinguish two extremum cases of insolation conditions, namely perfectly clear conditions (i.e., no variability) and heavily overcast conditions (i.e., again, no variability), contrary to other metric proposed such as $\sigma(Kt)$ [5, 16]. Nominal variability $\sigma(\Delta K^*(t, \Delta t)$ is a metric such a measure of the variability of the dimensionless clear sky index Kt^*.

3. Taylor power law, a statistical perspective in solar energy

3.1 Definition of the Taylor power law

Many fields exhibit complex process such as biology, ecology and, engineering sciences. The analysis of these complex process exhibited the universality of the Taylor power law defined by [17] by a scaling relationship more precisely described as" temporal fluctuation scaling" [18]. The Taylor power law (or temporal fluctuations scaling), is a scaling relationship between the standard deviation

$\sigma = \sqrt{\frac{1}{N}\sum_{i=1}^{N}(x_i(t) - <x>)^2}$ of a signal $x(t)$ and its mean value $<x> = \frac{1}{N}\sum_{i=1}^{N}x_i(t)$ estimated over a sequence of length N of the considered signal x(t) defined as in [19] and described by equation Eq.(4):

$$\sigma_{\Delta t} = C_0 <x>^{\lambda} \tag{4}$$

with $<.>$ defining the statistical average, and Δt is the increment corresponding to the time scales explored, C_0 is a constant and λ the Taylor exponent. The Taylor law is therefore a power law and a scaling relationship between the standard deviation of a phenomenon and its mean value.

3.2 Taylor law in solar energy data

Solar energy is a complex process. Particularly for insular context, this energy resource exhibits high fluctuations at all temporal and spatial short time scales. The analysis of the stochastic nature of this resource is in growing in the literature and have shown evidence of scaling properties despite its complexity [3, 5, 6, 20]. In this paper, an analysis of scaling properties of irradiance fluctuations is proposed. By analogy with Taylor law performed on several complex processes, we investigate in the study of Taylor power law performed on the intradaily variability of irradiance field, specifically on the $|\Delta K^*(t, \Delta t)|$. The metric $|\Delta K^*(t, \Delta t)|$ exposes directly the fluctuations' magnitude. Thus, we verify a scaling relationship between the nominal variability $\sigma_{|\Delta K^*(t,\Delta t)|}$ and the mean value $\mu_{|\Delta K^*(t,\Delta t)|}$. Therefore, the process of intradaily variability irradiance will obey power Taylor law if the equation Eq. (5) is verified:

$$\sigma(|\Delta K^*(t, \Delta t)|) = C_0\mu\left(\left|\Delta K^*(t, \Delta t)\right|\right)^{\lambda} \tag{5}$$

with λ for a given time scale Δt.

The four sites previously mentioned, characterized by tropical insular context hence exhibiting high variability, were chosen to test the consistency of this temporal fluctuation scaling method.

3.3 Criterion of the temporal limit of Δt

The time increment Δt or resolution of the irradiance data is a parameter that affects the magnitude. Moreover, the increment affects the length of daily data sampling. This is an important parameter to consider. In order to justify the choice of Δt threshold for our study, the Pearson coefficient is assessed between $log(\sigma(|\Delta K^*(t, \Delta t)|))$ and $log(\mu(|\Delta K^*(t, \Delta t)|))$ as a function of Δt. The threshold of Δt is considered as being the value of Δt when the Pearson coefficient is lower than 0.6. The Pearson coefficient is analyzed for several data sampling and for each site under the study. The results are exposed in **Figure 4**. According to the **Figure 4**, the Pearson coefficient is lower than 0.6 from $\Delta t = 3h$ for Tampon site but increases for $\Delta t = 4h$. We consequently considered that the threshold of Δt for our study would be $\Delta t = 4h$ which corresponds to a differentiation on a quasi half day since the daylight sequences are from 7 am to 5 pm (10 hours). Moreover, taking account of the length of daily data for Δt higher than 4 hours, we can deduce that is not representative for our study. Indeed, an average or a standard deviation on a small length of time series is not representative for our study and can give absurd results for our analysis of intradaily variability.

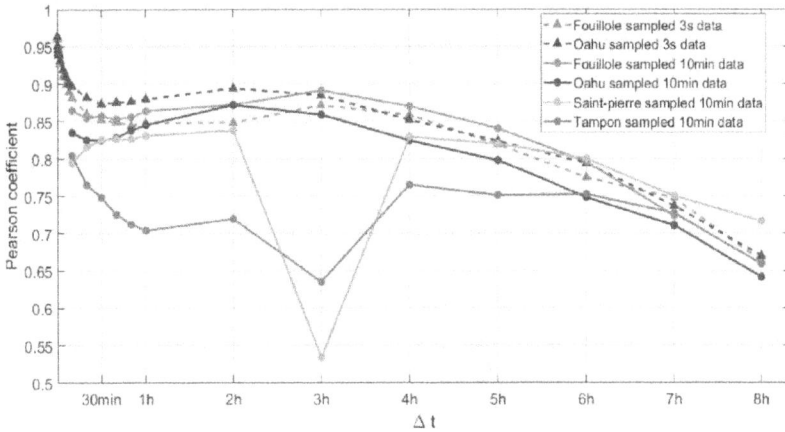

Figure 4.
Evolution of Pearson coefficient between $\log\left(\sigma(|\Delta K^*(t, \Delta t)|)\right)$ *and* $\log\left(\mu(|\Delta K^*(t, \Delta t)|)\right)$ *as a function of* Δt *conditionned to the data sampling.*

4. Verification of the existence of power Taylor law

4.1 Verification of the existence of power Taylor law for very short time scales dataset

The existence of power Taylor law is first verified for sampled data at very short time scales, i.e. 3 s, for the whole of dataset (2 years). This time scale of data sampling is available for Fouillole and Oahu measurement sites. The increment Δt ranges from 3 s until to the limit of 4 h as previously mentioned. The normalization of $\sigma(|\Delta K^*(t, \Delta t)|)$ by C_0 was done to remove the influence of specificities due to locations. Here, the goal is to highlight the existence of Taylor's law on irradiance data. **Figure 5** illustrates the evolution of the variance as a function of the average in log–log scale for an example of time scale $\Delta t = 30s$. We observe the existence of a power law between $\sigma(|\Delta K^*(t, \Delta t)|)$ an $\mu(|\Delta K^*(t, \Delta t)|)$ represented by weighted

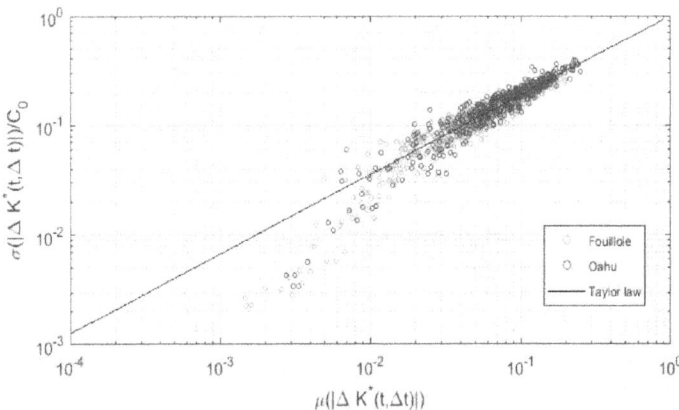

Figure 5.
Evolution of $\frac{\sigma}{C_0}$ *versus the mean* μ *for several* $\Delta t = 30s$ *with 3s sampled data.*

Δt	$\lambda_{Fouillole}$	λ_{Oahu}	Δt	$\lambda_{Fouillole}$	λ_{Oahu}
3 s	0.78	0.84	10mn	0.62	0.68
15 s	0.74	0.74	20mn	0.57	0.68
30s	0.71	0.70	30mn	0.55	0.73
45 s	0.70	0.69	40mn	0.54	0.76
1mn	0.68	0.68	50mn	0.55	0.79
2mn	0.65	0.67	1 h	0.55	0.83
4mn	0.62	0.68	2 h	0.56	0.85
6mn	0.60	0.69	3 h	0.61	1.02
8mn	0.58	0.68	4 h	0.68	1.10

Table 2.
Table of evolution of λ as a function of Δt.

least squares function [21, 22] in log–log scale plot. This power law is in accordance with the Taylor power law. The results showed that the process of intradaily variability irradiance obeys power Taylor law for sampled 3 s dataset and each Δt from $\Delta t = 3s$ to $\Delta t = 4h$ **Table 2** presents the values of λ as a function of time scales Δt for the two sites. The exponent λ of power Taylor law varies between 0.4 and 0.8 for $\Delta t = 30s$ to $\Delta t = 4h$. The majority of values are higher than 0.5. According to [23], Taylor's fluctuation scaling results from the ubiquitous second law of thermodynamics called the maximum entropy principle and the number of states, a concept borrowed from physics.

This power Taylor law verifies that:

$$\frac{\sigma(|\Delta K^*(t, \Delta t)|)}{C_0} = \mu(|\Delta K^*(t, \Delta t)|)^{\lambda} \tag{6}$$

4.2 Verification of the existence of power Taylor law for short time scales dataset

The temporal fluctuation scaling is analyzed by assessing power Taylor law on 10 min sampled data which is available on the whole of sites under study (Tampon, Saint-Pierre, Fouillole, Oahu).

The result of this analysis for $\Delta t = 1h$ over all measurements sites is presented in **Figure 6**. In **Table 3**, the results are described for each Δt and each measurement sites. Two comments can be given. Firstly, the results showed the consistency of power Taylor law for 10 min sampled data for every increment ($\Delta t = 10\,min$ from $\Delta t = 4h$). The result for $\Delta t = 1h$ and each measurements site is illustrated in **Figure 6**. Secondly, the λ coefficients exhibit a different trend from a site to another with particularity for Saint-Pierre site presenting the lowest values of this coefficient (λ lower than 0.5 for each Δt). This can highlight a particularity of factors governing the process of intradaily variability irradiance at this measurement site clearly different from the other locations.

The same analysis is also done for sampled data at 30s, 1 min, 2 min, 5 min and showed the consistency of power Taylor law for intradaily variability irradiance process. The results are presented in Section 5, **Figure 7** for the study of the evolution of λ as a function of data sampling.

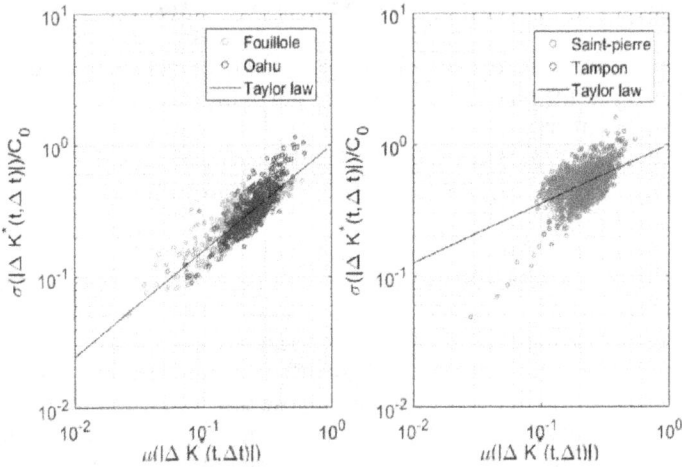

Figure 6.
Evolution of $\frac{\sigma}{C_0}$ versus the mean μ for $\Delta t = 1h$ for 10 min sampled data.

Δt	$\lambda_{Fouillole}$	λ_{Oahu}	$\lambda_{Saint-pierre}$	λ_{Tampon}
10 min	0.70	0.80	0.41	0.63
20 min	0.71	0.84	0.36	0.63
30 min	0.72	0.89	0.33	0.64
40 min	0.71	0.93	0.31	0.67
50 min	0.72	0.97	0.29	0.71
1 h	0.72	1.02	0.26	0.74
2 h	0.74	1.16	0.30	0.81
3 h	0.77	1.14	0.36	0.76
4 h	0.78	1.12	0.31	0.68

Table 3.
Table of evolution of λ as a function of Δt.

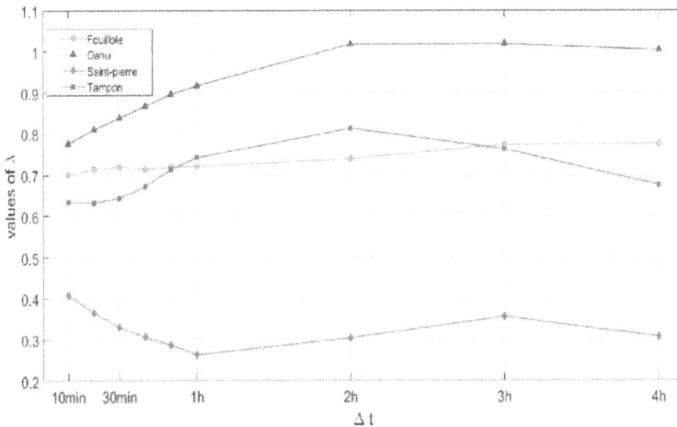

Figure 7.
Evolution of λ as a function of Δt conditioned to the data sampling.

5. Illustration of λ as a function of temporal parameters

This analysis allows assessing if there is a dependence between λ and temporal parameters such as the increment Δt and the time scale data sampling, conditioned to the measurement sites.

5.1 Evolution of coefficient λ as a function of increment Δt parameter conditioned to data sampling

The first study assessing the evolution of λ as a function of Δt showed that λ coefficients increase until to about $\Delta t = 2h$ excepted for Saint-Pierre site where λ coefficients decrease until to $\Delta t = 1h$. Moreover, it is observed that λ coefficients become quasi constant from $\Delta t = 2h$ (**Figure 8**). The time increment is a parameter that affects the magnitude, hence we can suppose that from $\Delta t = 2h$ until to $\Delta t = 4h$ the consistency of λ coefficient characterizes ramp rate as being quasi invariant. Moreover, the increment affecting the length of data sampling can alter the accuracy of λ value.

The profile of evolution of coefficients λ as a function of Δt does not vary a lot from a time scale of sampling to an other (**Figure 7**). This highlights the non dependence of coefficients λ to time scale of data sampling. Consequently, there is a consistency of evolution trend of λ as a function of Δt, in particular, an averaged trend of λ whatever the data sampling available but specific to a site. Thus, synthetic time series data at high frequency which are not commonly available would be produced from lower frequency by using nominal variability modeling from power Taylor law. This may be useful for inefficient forecasting model at very short time scale for example Numerical Weather prediction (NWP) models such as in [6].

5.2 Verification of the Taylor law stationarity

The evolution of coefficients λ as a function of Δt is assessed for several years available in our data set for Fouillole and Oahu. This coefficient λ is computed for a database of two years. This analysis is performed for several data sampling. **Figure 9** represents the results. From a year to another, the profile is substantially the same. This highlights the yearly stationarity of the evolution of λ as a function of Δt. We can deduce that for an analysis of λ, the user needs only one year data set to generalize his results. Nevertheless, to support this result and extend this analysis, more available years data are needed.

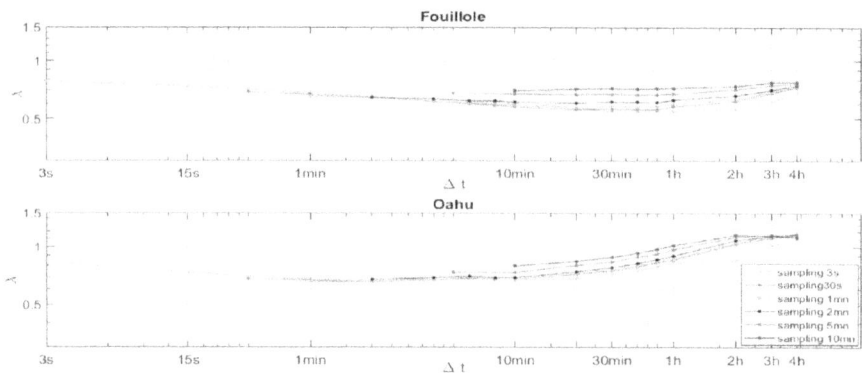

Figure 8.
Evolution of λ as a function of Δt.

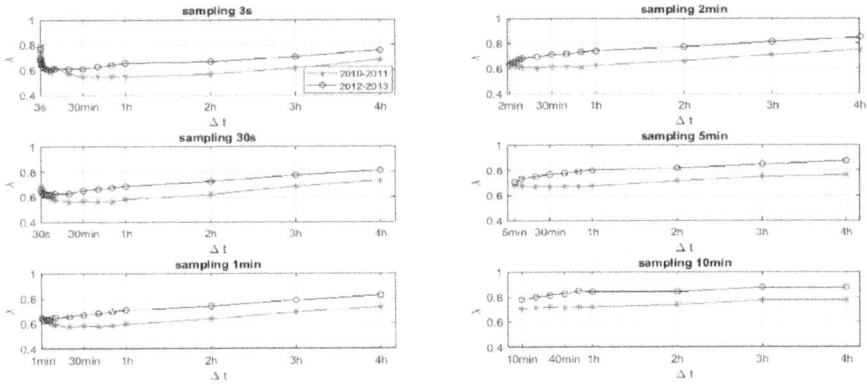

Figure 9.
Evolution of λ as a function of Δt conditioned to the data sampling.

6. Temporal fluctuations scaling analysis for PV power output

6.1 Characteristics of photovoltaics panels and PV power output modeling

In order to verify the consistency of Taylor power law for PV power output (power production from photovoltaic panel), the PV power output time series is simulated and obtained by a theoretical model for a first approach. The PV power output modeling is calculated by the following equation Eq. (7) such described in [24]. We have chosen arbitrarily a classic panel of monocrystalline technology for the simulation. The characteristics of the photovoltaic panel are described in **Table 4**. The required parameters for this modeling are the number of panels set at 1, the panel area, and the panel's efficiency according to the theoretical model equation Eq. (7).

$$P_s = N_P * GHI * A_P * \eta_P \qquad (7)$$

where, GHI is the measured irradiance in $W.m^{-2}$, A_P is the panel area, N_P is the number of panels, and η_P is the panels' efficiency.

Technology	Monocrystalline
Nominal power	$P_c = 185W_c$
Voltage for maximal power	$V_{MPP} = 33.7V$
Current for maximal power	$I_{MPP} = 5.49A$
Voltage of open circuit	$V_{oc} = 40V$
Current of short circuit	$I_{sc} = 5.8A$
Dimension of photovoltaic cells (mm) A_P	$125 * 125$
Number of cells	$5 * 10(50)$
PV module dimension (mm)	$1600 * 900 * 35$
Panels' efficiency	$\eta_P = 15\%$

Table 4.
Characteristics of photovoltaic panel parameters.

The data in **Table 4** are based on measurements under the standards conditions SRC (Standard Reporting Conditions, knowledge also: STC or Standard Test Conditions) which: an illumination of $1\,kW/m^2$ (1 sun) to a spectrum AM 1.5; a temperature of cell of 25°C.

The aim here is to obtain an output power profile to evaluate the existence of the Taylor power law. Considering the transfer function between the GHI and the power output of the panel, one should expect the same results found for irradiance. We decided on a first approach to verify Taylor's law on simulated data which should be a good approximation of the real case. To reinforce this study in perspective, we will need real data from PV power output. An example of a sequence of PV power output time series is presented in **Figure 10**.

The stochastic component of PV power output is obtained by removing the solar-geometry effects. Similarly to the clear sky index Kt^* (ratio of measured GHI to theoretical clear sky GHI) defined as Eq. (1), the detrending of PV output is described by the equation Eq. (8).

$$P* = \frac{P_m}{P_{clear}} \tag{8}$$

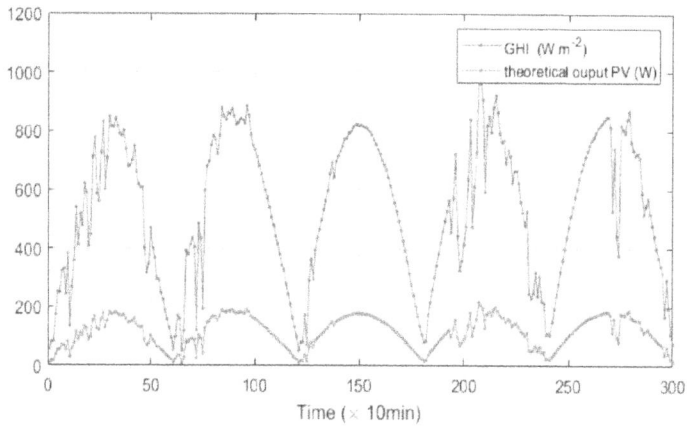

Figure 10.
Time series of GHI and the theoretical ouput PV power corresponding.

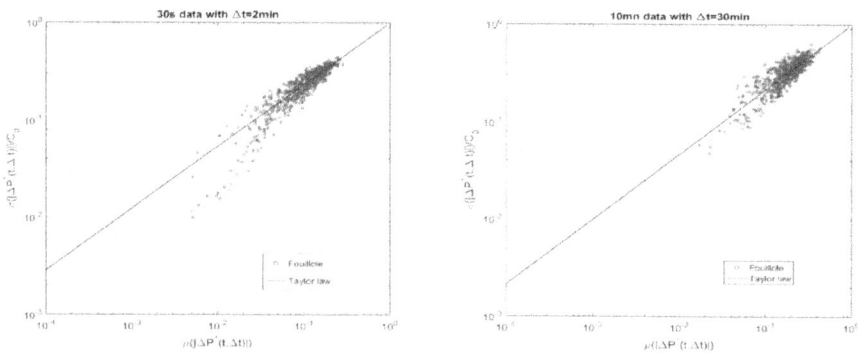

Figure 11.
Evolution of $\frac{\sigma}{C_0}$ versus the mean μ for $\Delta t = 2\,min$ for 30s sampled data and $\Delta t = 30\,min$ for 10min sampled data.

where P_m is the PV output estimated from measured irradiance for index m and index *clear* refers to PV output estimated from theoretical clear sky irradiance. For this analysis, we used the data from Fouillole measurement site. In [5], a metric called power variability is defined by $\sigma(\Delta P_{\Delta t})$. As the previous study, we define the metric of the ramp rate standard deviation from PV output power by this equation Eq. (9).

$$\sigma\left(\Delta P^*(t, \Delta t)\right) = \sqrt{Var[\Delta P^*(t, \Delta t)]} \tag{9}$$

By analogy with power Taylor law performed for irradiance, we verify a scaling relationship between $\sigma_{|\Delta P^*(t, \Delta t)|}$ and the mean value $\mu_{|P^*(t, \Delta t)|}$ for several increments from Δt (**Figure 11**).

6.2 Power Taylor law consistency for PV ouput area

The power Taylor law for PV power output has been verified for very short time scale (30s sampled data) and short time scale (10 min sampled data). The results showed a consistency of Taylor power law for PV area output (**Figure 12**) which is an expected result due to the relation between irradiance and PV power output modeling. Therefore, there is no changing of the inherent cause of variability. The results have shown evidence for the existence of temporal fluctuation scaling for PV power output data. Hence, the ramp rate standard deviation of power PV can be modelized by this equation Eq. (10):

$$\sigma(|\Delta P^*(t, \Delta t)|) = C_0 \mu^\lambda_{|\Delta P^*(t, \Delta t)|} \tag{10}$$

The λ power coefficients show significant similarities both in the values and in the evolution profile of the λ as a function of increment Δt between power PV field and irradiance field. We can deduce from this first approach that irradiance data are sufficient to model the ramp rate standard deviation of PV ooutput by power Taylor law without having to use P^*. Hence, the modeling ramp rate standard deviation of PV output can be described by this following equation:

$$\sigma|\Delta P^*(t, \Delta t)| = C_0 \mu\left(|\Delta P^*(t, \Delta t)^\lambda\right) \tag{11}$$

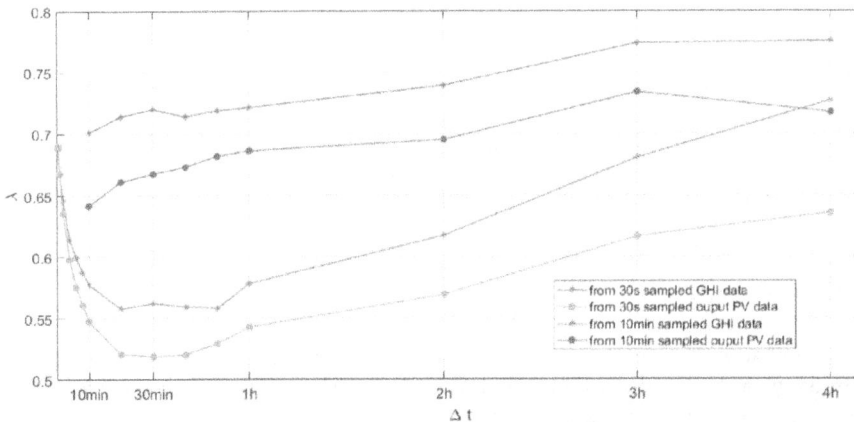

Figure 12.
Evolution of λ as a function of Δt conditioned to the data sampling from GHI data and from estimated PV power ouput data.

Consequently, theoretically, the user does not need to have available PV output data set to characterize ramp rate standard deviation of PV output. To reinforce this study in perspective, we will need real data from PV output.

7. Discussion

The development of installed photovoltaic (PV) power increases problems related to the underlying variability of PV power production. Characterizing the underlying spatiotemporal volatility of solar radiation is a key ingredient to the successful outlining and stable operation of future power grids [25]. In literature, scientifics attention and studies related to the understanding of weather-induced PV power output variability are in full development.

Each time scale interval of solar generation is associated with a specific problem of load management challenges. In [5], a characterization of how solar energy's resource variability impacts energy systems and a definition of the temporal or the spatial scales context are given. In our study context that concerns very short time scales fluctuations, voltage control issues are a specific problem [5, 26]. This observation implies an understanding of the ramp's variance at very short time scales.

As PV power variability is mainly determined by irradiance variability, irradiance variability quantifications are essential to the successful outlining and stable operation of future power grids [27]. Variability in irradiance itself as interesting as variability in irradiance increments. Indeed, irradiance increments are transitions from one point in time to another, namely ramp rates. Irradiance variability and irradiance increments impact the system PV differently. Irradiance variability mainly impacts a PV system's yield and the proper dimensioning of energy storage, while increment variability affects power quality as well as the maintenance of the generation load balance [25]. Therefore, our study is firstly focused on increment variability in irradiance. Then, this analysis is applied to PV power output time series.

The works in this article bring a complementary understanding of underlying variability. The results highlighted a new modelization of ramp rate's variance of irradiance and PV power output based on the fluctuations' magnitude from Taylor power law. This model makes it possible to extrapolate the resulting variability of PV power output. Moreover, a synthetic time series data at high frequency which are not commonly available would be produced from lower frequency by using nominal variability modeling from power Taylor law. This new model fills a gap in temporal scales. This may be useful for inefficient PV power output and irradiance forecasting model at very short time scale for example Numerical Weather prediction (NWP) models.

Analysis of λ exponent has shown that the user needs only one year data set to generalize his results. Nevertheless, to support this result and extend this analysis, more available years of data are needed.

8. Conclusion

This chapter presented a characterization of the irradiance and PV power output intradaily variability describing a temporal fluctuation scaling. By analogy of environmental complex process, the works have demonstrated that power Taylor law is verified for the ramp rate's variance of irradiance named nominal variability, namely the standard deviation of the changes in the clear sky index $\sigma(\Delta | P^* (t, \Delta t))$ even for very short time scales. Hence, this study allowed to model this metric by a power law based on $\mu(\Delta P^*(t, \Delta t))$. The exponent seems to depend on the location of sites. This can be due to different factors causing cloudy formation specific to a site which is the source of the ramp rate's variance particularities. The invariance of λ evolution profile as a function of Δt conditioned to the sampling of data highlighting the possibility to approximately model ramp rate variance at high frequency from lower frequency data. The stationarity of λ evolution profile as a function of Δt from a year to another showed that the user does not need a long dataset to establish this power law describing ramp rate's variance. Moreover, the study showed evidence that this modeling also applies to PV power output. For all increments Δt of this study from 30s sampled and 10 min sampled data, the exponent values of Taylor power law λ are between 0.5 and 0.8. The results of these works are a statistical perspective in PV power output area and introduce the multifractility analysis of intradaily variability PV power output which is a prerequisite of this characterization, enabling its high penetration.

Conflict of interest

The authors declare no conflict of interest.

Thanks

The authors thank Laboratory PIMENT (Laboratoire de Physique et Ingénierie Mathématique pour l'Energie et l'environnement) from University of La Réunion, for providing ground measurements databases at locations Tampon and Saint-Pierre and NREL (National Renewable Energy Laboratory) website for providing ground measurements databases at locations Oahu.

Author details

Maina André and Rudy Calif*
EA 4539 LaRGE (Laboratoire de Recherche en G'eosciences et Énergies), Université des Antilles, Pointe-á-Pitre, France

*Address all correspondence to: rudy.calif@univ-ag.fr

IntechOpen

References

[1] Lenox C, Nelson L. Variability Comparison of Large-Scale Photovoltaic Systems across Diverse Geographic Climates. In: 25th European Photovoltaic Solar Energy Conference, 2010,Valencia, Spain.

[2] Perez R, Kivalov S, Schlemmer J, Hemker K, Hoff T. Parameterization of site-specific short-term irradiance variability. Solar Energy. 2011;85: 1343-1353

[3] Stein JS, Hansen CW, Reno MJ. The Variability Index: A New and Novel Metric for Quantifying Irradiance and PV Output Variability. In: World Renewable Energy Forum, 2012, Denver, Colorado.

[4] Kang BO, Tam KS. A new characterization and classification method for daily sky conditions based on ground-based solar irradiance measurement data. Solar Energy. 2013; 94: 102–118

[5] Perez R, David M, Hoff TE, Jamaly M, Kivalov S, Kleissl J, Lauret P and Perez M. Spatial and temporal Variability of solar energy. Foundations and Trends® in Renewable Energy. 2016;1(1):1-44

[6] Lauret P, Perez R, Mazorra Aguiar L, Tapachès E, Diagne HM, David M. Characterization of the intraday variability regime of solar irradiation of climatically distinct locations. Solar Energy. 2016; 125: 99-110

[7] Hansen T and Phillip D. Utility solar generation valuation methods. Tecnichal report.

[8] Marcos J, Marroyo L, Lorenzo E, Alvira D, Izco E. Power output fluctuations in large scale PV plants: One year observations with one second resolution and a derived analytic model. Prog. Photovolt. Res. Appl. 2011; 19: 218–227

[9] Mills A, Ahlstrom M, Brower M, Ellis A, George R, Hoff T, Kroposki B, Lenox C, Nicholas M, Stein J, Wan YW. Understanding Variability and Uncertainty of Photovoltaics for Integration with the Electric Power System. IEEE Power and Energy Magazine. 2011; 9(3)

[10] Van Haaren R, Morjaria M, Fthenakis V. Empirical assessment of short-term variability from utility-scale solar PV plants. Prog. Photovol. Res. Appl. 2012;22: 548–559

[11] Lauret P, Voyant C, Soubdhan T, David M, Poggi P. A benchmarking of machine learning techniques for solar radiation forecasting in an insular context. Solar Energy. 2015; 112: 446-457.

[12] Diagne HM. Gestion intelligente du réseau électrique Réunionnais. Prévision de la ressource solaire en milieu insulaire. PhD thesis, Université de La Réunion; 2015.

[13] Bird RE, Hulstrom RL. Simplified the Clear Sky Model for Direct and Diffuse Insolation on Horizontal Surfaces. In: Technical Report No. 1981; Solar Energy Research Institute.

[14] Ineichen P. Comparison of eight clear sky broadband models against 16 independent data banks. Solar Energy. 2006; 80: 468–478

[15] Kasten F. Parametrisierung der globaslstrahlung durch bedekungsgrad undtrubungsfaktor. Ann Meteorol 1984; 20:49-50.

[16] Skartveit A, Olseth JA. The probability density of autocorrelation of short-term global and beam irradiance. Solar Energy. 1992; 46(9):477-488

[17] Taylor LR. Aggregation, variance and the mean. Nature. 1961; 189: 732-735

[18] Eisler Z, Kertész J. Scaling theory of temporal correlations and size-dependent fluctuations in the traded value of stocks. Physical Review E. 2006; 73(040109)

[19] Calif R, Schmitt FG. Modeling of atmospheric wind speed sequence using a lognormal continuous stochastic equation. Journal of Wind Engineering and Industrial Aerodynamics. 2012; 109: 1-8

[20] Lave M, Reno MJ, Broderick, RJ. Characterizing local high-frequency solar variability and its impact to distribution studies. Solar Energy. 2015; 118: 327-337.

[21] Street JO, Carroll RJ, Ruppert D. a note on computing robust regression estimates via iteratively reweighted least squares. The American Statistician. 1988; 42: 152-154.

[22] Draper NR, Smith H. Applied regression analysis, 3rd edition: Wiley series in probability and statistics; 1998.

[23] Fronczak A, Fronczak P. Origins of Taylor's power law for fluctuation scaling in complex systems. Physical Review E. 2010; 81(6).

[24] Khaled U, Eltamaly A M, Beroual A. Optimal power flow using particle swarm optimization of renewable hybrid distributed generation. Energies. 2017; 10 (1013)

[25] Lohmann, Gerald M. Irradiance Variability quantification and small-scale averaging in space and time: A short review. Atmosphere. 2018;9(7): 264

[26] Widen J, Wäckelgård E, Paatero J, Lund P. Impacts of distributed photovoltaics on network voltages: Stochastic simulations of three Swedish low-voltage distribution grids. Electric Power Systems Research. 2010;80(12): 1562-1571

[27] Mills A, Ahlstrom M, Brower M, Ellis A, George R, Hoff T, Kroposki B, Lenox C, Miller N, Milligan M, Stein J, Wan Y-h. Dark Shadows. IEEE Power and Energy Magazine. 2011;9(3):33-41

Solar Photovoltaic Technologies and Applications

Assessing the Impact of Spectral Irradiance on the Performance of Different Photovoltaic Technologies

Mohammad Aminul Islam, Nabilah M. Kassim,
Ammar Ahmed Alkahtani and Nowshad Amin

Abstract

The performance of photovoltaic (PV) solar cells is influenced by solar irradiance as well as temperature. Particularly, the average photon energy of the solar spectrum is different for low and high light intensity, which influences the photocurrent generation by the PV cells. Even if the irradiance level and the operating temperature remain constant, the efficiency will still depend on the technological parameters of the PV cell, which in turn depends on the used PV material's absorption quality and the spectral responsivity and cell structure. This study is devoted to the review of different commercially available technologies of PV cells include crystalline silicon (c-Si), polycrystalline silicon (pc-Si), cadmium telluride (CdTe), and copper indium gallium selenide (CIGS). We tried to correlate the spectral response or the photocurrent of different PV cells with the variations of the solar spectrum, environmental conditions, and the material properties and construction of PV cells.

Keywords: photovoltaics technology, average photon energy, spectral irradiance, spectral effect, photovoltaic performance

1. Introduction

The energy demand is increasing concurrently with the increase of the world's population and meeting the increasing energy demands including managing social, economic, and ultimate environmental issues are one of the greatest challenges of the present time. Solar energy, as one of the promising renewable energy sources, is becoming an important source of energy all over the world. Its huge development potential has attracted a lot of attention and the photovoltaic (PV) industry has been experiencing a large-scale development to replace traditional energy. Also, a significant increase in energy conversion efficiency and the decrease in the price of the solar panels along with various national policies over the world enhanced the solar PV-based energy generation with the least Levelized-cost-of-energy (LCOE). However, for getting optimum output, proper resource estimation is necessary to assess the feasibility of solar PV systems in any area. The output of any PV system's output indeed depends on the weather of its surroundings will be elaborated in this chapter.

IntechOpen

In contrast, there are several types including different technologies of photovoltaic modules in the market. However, the technological choices are very critical in the sense of the lack of guide and forecasting tools suited to the climates and environment of the installation sites. There have been many PV system projects going to fail due to the bad choice of PV technology where failure causes are influenced by the environmental parameters, such as heat, humidity, shadow, and dust, etc. Manufacturers provide a characteristic of PV modules measured in standard test conditions (STC), however, the performance cannot reach that level in real operating conditions. Besides, the PV module's performances and aging strongly depend on the climate and the surrounding environment of the installation site.

The investigation of PV performance under real external conditions became an important factor as a result of increasing trends of PV capacity over the world. Particularly, the performance of the PV module influence by the number of different external issues, such as, (i) spectral irradiance, i.e., the wavelength of incident light and light intensity, the efficiency of PV certainly varied with the variations in the spectrum of sunlight [1–4] and light intensity directly affect the short circuit current [5]; (ii) reflectivity of the module surface, however, the reflectivity that occurred in the module surface depends on the angle between the module surface and the incident angle [6]; and (iii) module temperature, particularly temperature of the module surface increases to 60-80 °C at noon and cause of the reduction in open-circuit voltage which also depend on the light intensity and airflow [7]. Thus, in each PV field, the factors that contribute to solar cell efficiency are different and the important considerations applied in each area are different. On the other hand, some types of PV modules show the degradation of power conversion efficiency under the long-term light exposer in the field and/or elevated temperatures. Particularly, due to the above-mentioned effects, the module efficiency and/or electrical parameters are observed to deviate from the nameplate value measured under Standard Test Conditions (STC) [8] in the real external condition. Besides, there are some other causes for which the energy production capability of a PV module is affected, such as installation angle; possible shadow, dust or snow deposition, etc. However, these mostly depend more on the details of the installation, not inherent to the module type and the physical properties of the module. It could be mention that the power output could vary as an impact of the above-listed causes while different types of the module installed in the same way; alternatively, similar types of module generate different power output due to the installation in a different way or different places. The variation of PV performance has been investigated by several authors in terms of geographical variability and technology. Some authors only focused on the effect of solar irradiation while other authors consider some of the above-mentioned factors. In this study, we also only reviewed the study that focuses on the effect of solar irradiance on the different PV technology.

2. Spectrum irradiance on earth surface

Solar irradiance on different locations of the earth is shown in **Figure 1** [9]. The maps highlight the global horizontal irradiation (GHI) which means that the overall irradiance from the sun reaches the earth's horizontal surface. It is related to the diffuse horizontal irradiance (DHI) and direct normal irradiance (DNI) as follows [10],

$$GHI = DHI + DNI \times \cos\theta \tag{1}$$

Where θ is the solar zenith angle. Areas with a high proportion of GHI include South-East China, Northern Europe, and the tropical belt around the equator.

GLOBAL HORIZONTAL IRRADIATION

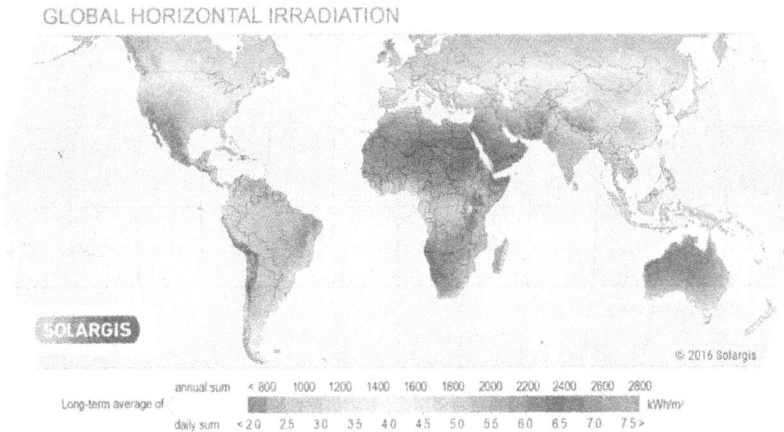

Figure 1.
Global horizontal irradiation (GHI) over the world [9].

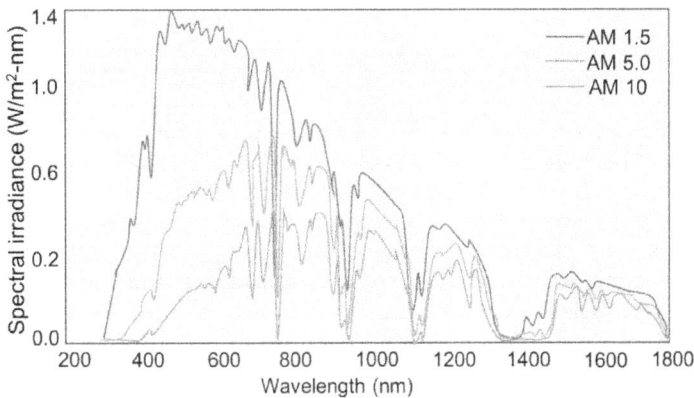

Figure 2.
Impact on the direct spectral irradiance of air mass (AM) simulated with the SMARTS model [11].

Since the electrical performance of PV devices is greatly affected by the incident light spectrum, hence, significant efforts have given by the PV community to develop methods and evaluate the impact of the spectral variations on the PV device performance over the last three decades. The parameters that have the highest impact on the spectrum distribution as well as on the PV module performance are (i) the air mass (AM), (ii) the perceptible water (PW), and (iii) the aerosols optical depth (AOD) [11].

The AM is a measure of the atmospheric absorption that affects the spectral content and the intensity of the solar radiation coming to the earth's surface. The impact of AM on the solar spectral distribution is shown in **Figure 2** [11]. Particularly, the solar spectral distribution just above the Earth's atmosphere-in the relative vacuum of space is commonly referred to as an air mass zero (AM0). And the AM = 1.0 at sea level when the sun is directly overhead (zenith angle, $\theta z = 0$). As the θz increases, the path passes by the sun spectrum through the atmosphere become longer, and AM increases. The AM could estimate simply using a trigonometric function of the zenith angle:

$$AM = \frac{1}{\cos \theta_z} = sec\, \theta_z \tag{2}$$

where θz is the angle of incidence or solar zenith angle.

The above equation is quite accurate for θz ≤80 degrees, however, more complex and precise models are necessary when the sun goes near the horizon. Moreover, the distribution of the outdoor solar spectrum varied during the day due to the presence of water vapors and aerosol in the air. Thus, the real spectrum at the earth's surface is infrequent to fit with the AM1.5 standard solar spectrum as defined in standard IEC 60904–3 and/or ASTM GE173–03 [12]. Specifically, the spectral power distribution observed in the sun at an angle of about 48.2° is specified as AM 1.5 spectrums (as in **Figure 3(a)**). The power density of AM1.5 light is about 1,000 W/m². The standard AM 1.5 spectrum is known as solar constant and is normally used in solar cell analysis. **Figure 3(b)** shows the spectral distribution of sunlight under the different air masses.

Another important parameter that needs to be considered for understanding solar irradiance on the earth's surface is the clearness index (K$_T$). Particularly, K$_T$ is defined by the ratios of the solar radiation for a particular day and the extraterrestrial solar radiation for that day. It could also be defined by hourly as shown below:

$$K_T = \frac{H}{H_o} (daily) \ and \ k_T = \frac{I}{I_o} (hourly) \tag{3}$$

Where H and I represent the total measured and Ho and Io are represent the extra-terrestrial solar radiation which could be calculated using several approaches [14]. This value of K$_T$ or k$_T$ lies between zero and one which contingent on atmospheric conditions. For clear sky conditions, K$_T$ is near 1 and if the sky is very cloudy and/or turbid and/or heavily overcast, K$_T$ becomes less than 0.4. Several laboratories have been developed computational models considering spectral direct beam during the clear sky and hemispherical diffused irradiances on a surface either horizontal or tilted condition for a certain location and time [15]. Other than the above parameters, the outdoor energy yield and performance of the PV modules further depend on a large number of on-site factors or local factors such as ambient temperature, wind, and rain. These undefined factors may also influence significantly amount of solar radiation that arrives on the surface of the PV module.

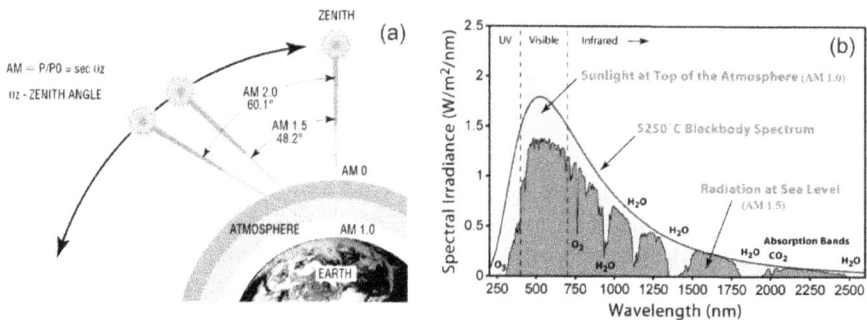

Figure 3.
(a) The path length (in units of air mass) changes with the zenith angle (b) spectral distribution of solar energy [13].

Certainly, it is important to analyze the influences of all the above on-site environmental factors on the outdoor performance of different types of PV modules, for finding out the best-suited technology for a specific location and enabling more widespread deployment.

3. PV performance parameters and spectrum

3.1 PV performance parameters

The electrical power generated in a solar cell or PV device can be modeled with a well-known equivalent circuit as shown in **Figure 4** which includes a shunt resistance parallel with a diode and a series resistance [16].

This equivalent circuit can be used for either an individual cell, a multi-cell module, or an array consisting of multiple modules. Using this model and considering constant temperature and solar radiation, the current–voltage equation for a solar cell or module could be expressed as shown in Eq. (4).

$$I = I_0 \left[\exp \left(\frac{qV}{A} \right) - 1 \right] - I_L + \frac{V - IR_s}{R_{sh}} \tag{4}$$

Where, I_L is the light generated current, Io is the dark saturation current, Rs is the series resistance, Rsh is the shunt resistance, A is the modified cell or module ideality factor that can be expressed as:

$$A \equiv \frac{N_s n_i KT}{q_e} \tag{5}$$

where Ns is the number of cells or modules that are connected in series, n_i is the diode ideality factor for a cell, K is the Boltzmann constant, q_e is the electron charge, T is the cell or module temperature.

Figure 5 shows the current–voltage (I-V) characteristic curves of a solar cell or a module. Particularly, the power generated by the solar cell or module is the product of the current (I_{mp}) and voltage (V_{mp}). It should be noted that five parameters, such as I_L, I_o, R_s, R_{sh}, and A, determine the current and voltage generated in a cell or module, thus the impact of external factors, such as solar radiation and temperature could be analyzed from the change of these values. In general, the FF is directly affected by series resistance, and it is found that the fill factor of a solar cell decreases by about 2.5% for each 0.1 Ω increase in series resistance [17]. On the

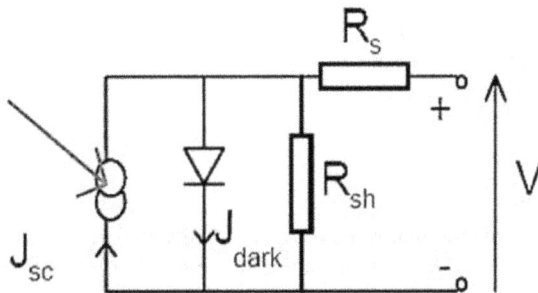

Figure 4.
Equivalent circuits for a solar cell in a single diode model, including series and shunt resistance [16].

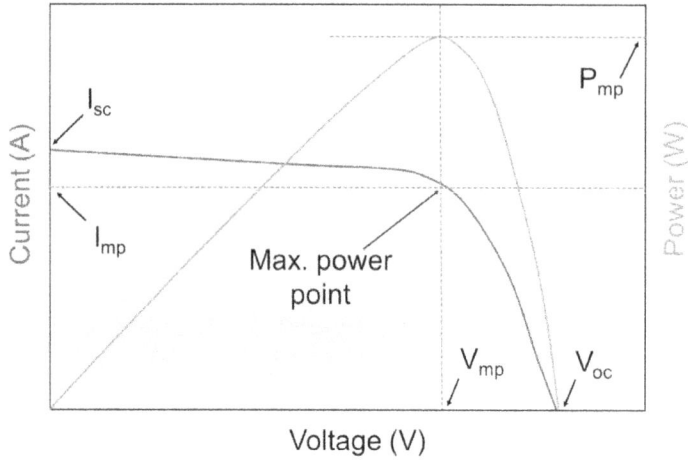

Figure 5.
Typical current–voltage (I-V) and power-voltage characteristic curves of a solar cell.

other hand, Rsh is reduced if the leakage current is increased in a solar cell. If there any light and temperature-activated defects available in a solar cell, then leakage current could be increased, alternatively Rsh could be reduced as the increase of irradiance intensity or temperature. Finally, FF and Voc will be reduced. For an ideal case, Rs = 0, Rsh = ∞ and n_i = 1, the open-circuit voltage, Voc could be expressed as,

$$V_{oc} \approx \frac{KT}{q_e} \ln\left(\frac{I_L + I_o}{I_o}\right) \tag{6}$$

For a very small applied voltage (V ≈ 0), the diode current, I_o is negligible or zero, then from Eq. (6), we can find,

$$I \approx I_L \approx I_{sc} \tag{7}$$

Where Isc is a short circuit current. Now Eq. (9) becomes,

$$V_{oc} \approx \frac{KT}{q_e} \ln\left(\frac{I_{sc} + I_o}{I_o}\right) \tag{8}$$

The Voc and Isc rectangle description as shown in **Figure 5** offers a useful means for characterizing the maximum power point [18]. The fill factor (FF) is defined as the ratio of the maximum power to the product of Voc and Isc and is less than one at all times. FF indicates the squareness of the I-V curves and can be defined from the ratios of two rectangles (**Figure 5**) as,

$$FF = \frac{P_{mp}}{I_{sc} V_{OC}} = \frac{I_{mp} V_{mp}}{I_{sc} V_{OC}} \tag{9}$$

Where P_{mp} denotes the maximum power of the solar cell or module, Imp and Vmp are the current and the voltage values at the maximum power point, respectively. Moreover, the most significant Figure of merit for a solar cell or PV module is its power conversion efficiency, η, which is specified as,

$$\eta = \frac{V_{OC}I_{SC}}{P_{in}}FF \qquad (10)$$

Where Pin denoted the power of incident light that is determined by the characteristics of the light spectrum incident onto the solar cell or PV module. The power of the incident light spectrum, Pin can be express as,

$$Pin = G \times A \qquad (11)$$

Where A is the surface area of the solar cell or PV module and G is the total spectral irradiance, which could be defined as [19],

$$G = \int_0^\infty \varepsilon_\lambda f(\lambda)d\lambda \qquad (12)$$

Where, $f(\lambda)$ is the flux density (number of incident photon per unit area and unit time) for a specific wavelength of the photon with energy, ε_λ and wavelength, λ.

3.2 Spectral response and quantum efficiency

Particularly, the light to the electrical power conversion efficiency of a solar cell or a module is an inherent property that depends on the type of semiconductor material and the manufacturing process. However, this efficiency also depends on the environment of the installation site, especially on the hours of equivalent peak spectral irradiance in a day and/or temperature. The PV module characteristics that we find in the nameplate are typically measured at standard testing conditions (STC), the irradiance of $1000Wm^{-2}$ at AM 1.5 and 25 °C of cell temperature. In fact, these conditions hardly exist because the outdoor spectrum is far different from the STC condition, which also varied by location and season. The response to the spectral variation by different types of PV modules vastly depends on its material properties and structure. This response is primarily determined by the bandgap of the materials used in fabrication, which sets the upper wavelength limit of the spectral response (SR). More specifically, SR is depending on the PV material's bandgap, cell thickness, and carrier transport mechanisms in the device. Secondly, device structure, means the position of the absorber material and other supporting layers has a significant effect on the spectral response. Also, the variation of electrical parameters of different types of PV module/device as an impact of various environmental factors depends on the technology (device structure and materials). On the whole, the PV device performance and SR is proportional for specific PV devices, where SR is defined as:

$$SR(\lambda) = J_L(\lambda)/G(\lambda) \qquad (13)$$

Where $J_L(\lambda)$ represents the light-generated current density for a specific wavelength "λ" and $G(\lambda)$ is the spectral irradiance of the incident light measured in W/m^2-nm. However, in state-of-the-art solar cell or PV modules, the spectral response is defined as the short-circuit current, Isc(λ), resulting from a single wavelength of light normalized by the maximum possible current [20–23].

$$SR(\lambda) = \frac{I_{SC}(\lambda)}{qAf(\lambda)} \qquad (14)$$

Where, q is the electronic charge 1.6 x 10–19 C, A is the surface area of the PV device and $f(\lambda)$ is the incident photon flux (number of photons incident per unit area per second per wavelength). Besides, the SR of the PV devices is also estimated in terms of quantum efficiency (QE), which indicates that how efficiently a PV device converts the incident light to a charge carrier that flows through the external circuit [24], details on QE has been discussed next section. In that case,

$$SR(\lambda) = QE(\lambda)\frac{q.\lambda}{h.c} \tag{15}$$

In the case of PV modules, J_L is approximately the same in value as the short-circuit current density (Jsc) [25]. Thus, with the help of the above equations, Jsc can be expressed as,

$$I_{sc}(\lambda) = \frac{q}{h.c}\int SR(\lambda).G(\lambda).\lambda.d\lambda \tag{16}$$

It could be seen in Eq. (16) that Jsc can be estimated by the SR for PV modules which certainly have prime importance in evaluating PV materials and device characteristics. Particularly, the degree to which the SR and the incident irradiance spectrum varies gives rise to a spectral effect on the device current and efficiency. The SR of different types of the module at AM1.5G spectrum (up to 1300 nm) is shown in **Figure 6** to confirm the response is different for different technologies [10]. As seen in Eq. (16), Isc is affected by the spectrum. Particularly, the spectrum variations are also influenced the other PV output parameters, viz. FF, Voc, and η. To determine the magnitudes of these effects on different technology-based PV devices, various performance review studies were carried out [26–31].

Particularly, The SR shows represent the current produced by a solar cell for per watt of irradiance at each wavelength of the photon. As seen in **Figure 7** that SR towards the higher wavelength region is lower because photons in this region have energy less than the material bandgap threshold. As a result, the effect of spectral variation on the output of PV devices is most pronounced in narrow SR technologies such as a-Si and CdTe. Especially narrowest SR is seen for the a-Si that is also discussed in the literature [32–35]. For simplification of SR and PV performance,

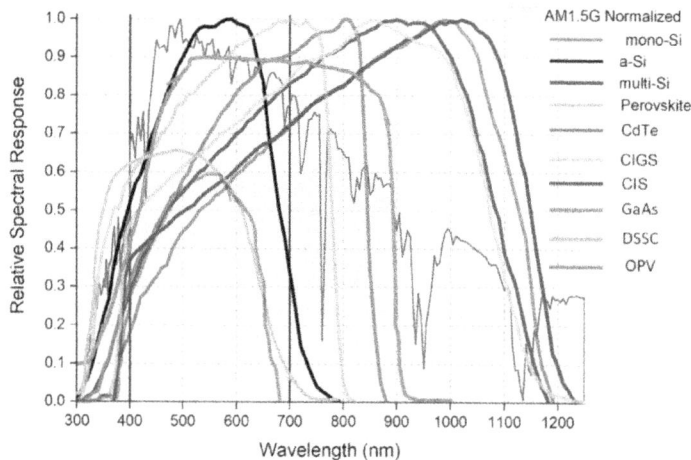

Figure 6.
Spectral response characteristics of different solar module technologies, modified from [26–35].

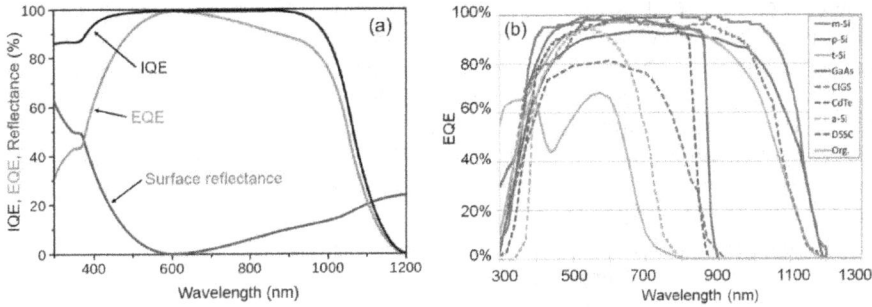

Figure 7.
(a) variation of EQE, IQE, and reflectance with the wavelength of a c-Si solar cell (collected and modified from Wikipedia), (b) EQE of different PV solar cell technology [41].

research is commonly used one-dimensional terms, such as spectral mismatch factor (MMF) [32–34, 36], the useful fraction (UF) [37], average photon energy (APE) [38, 39]. In the case of MMF and UF, their values should be a known factor for a specific module understudy, however, the SR data is not available publicly and analysis complexity arises. Besides, APE is denoted by the unit of an electron volt (eV) which signifies the average incoming photon energy. The equation for calculating is as follow:

$$APE = \frac{\int_{p\lambda}^{q\lambda} E(\lambda) d\lambda}{q_e \int_{p\lambda}^{q\lambda} f(\lambda) d\lambda} \tag{17}$$

Where, $E(\lambda)$ represents the energy of the incident photon and $f(\lambda)$ is the incident photon flux at wavelength λ, and p_λ and q_λ are the integration limits indicate the lower and higher absorption wavelength, which are 300 and 1200 nm as shown in **Figure 7**. Particularly, APE varies on a daily and seasonal basis due to the increase of air mass at sunrise and sunset compared to noon and in winter compared to summer. For example, when the sun is above the horizon, the spectral irradiance is red-shifted and the APE becomes low. APE rises again to a high around noon during the day. Moreover, the APE is higher in the summer months than in winter because the zenith angle of the sun is higher in summer. Besides, the atmospheric water, cloud cover, and/or aerosol content affect the APE due to light absorption and scattering. For most of the PV modules, the APE effect on performance seems to be linear. The spectral photon flux density denoted in joules can be expressed as below for a specific wavelength λ:

$$f(\lambda) = \frac{G(\lambda)}{E_\lambda} = \frac{G(\lambda)}{\frac{hc}{\lambda}} \tag{18}$$

where 'h' is the Planck constant and 'c' is the light velocity in vacuum.

The SR and QE are conceptually similar to each other. Particularly, SR is the ratio of the generated current in a solar cell per unit incident power, while QE denoted the ratio of the number of generated carriers and the number of the incident photon on the solar cell. In another way, the QE of a solar cell represents the amount of current the cell produces for a particular wavelength of an incident photon. Knowing the QE of a particular PV technology is important because by integrating QE for the whole solar spectrum, the current generation capability of PV solar cells could be realized. Interestingly, the QE value could exceed 100% for a PV solar cell in the case of multiple excitation and generation (MEG). In that case, one incident photon could generate several electron–hole pairs as an impact of multiple excitations. The MEG properties are typically seen in quantum-dot solar cells [40]. However, all the incident photons on the cell surface cannot be absorbed due to surface optical properties, such as absorption and reflection. Thus, QE is divided into two terms, (i) external QE (EQE) and (ii) internal QE (IQE) which simply differ by the photons reflection properties of a PV solar cell. In the case of EQE, all photons that impinge on the cell surface are taken into account, while in the case of IQE, only photons that are absorbed (not reflected) by the solar cell are considered. The graphical representation of EQE and IQE is shown in **Figure 7**.

High EQE is a precondition for high-power PV applications, which depends on the absorption coefficient of the absorber material of a PV solar cell, the carrier excitation quality, and carrier recombination rate or the amount of electron transport to the electrodes. The mentioned QE in Eq. (15) is typically EQE, which is directly related to the current generation by a solar cell [41]:

$$J_{sc} = q \int_0^\infty \varphi_\lambda(\lambda).EQE(\lambda).d\lambda \tag{19}$$

Where, with q is the charge of electron and $\varphi_\lambda(\lambda)$ the incident spectral flux density, indicating the incident number of photons of wavelength λ on the cell surface per unit of area, per unit of time and EQE could be defined as:

$$EQE = \frac{electron/s}{photon/s} = \frac{current/e}{total\ photon\ power/h\nu} \tag{20}$$

The relation between IQE and EQE could be defined as:

$$IQE = \frac{EQE}{1-L} = \frac{EQE}{1-R-T} \tag{21}$$

Where L is the total optical loss that occurred in a solar cell either through reflection or transmission or both. Particularly, for maximizing EQE, the optical loss should be minimized. To reduce the optical loss, anti-reflection coating, and back-reflection coating is applied in the current PV technologies.

3.3 Spectral irradiance and temperature

Solar irradiance and surface air temperature are two key factors for investigating the PV module performance. Particularly, the increase in solar irradiation is a cause of the increase in air temperature and vice versa. On the other hand, the increase in solar irradiance is proportionally increased the power output of the PV module, however, module output decrease with the increase of temperature [42]. Usually, the output and temperature of the PV modules are considered to be linear. The effect of temperature mostly depends on the absorber material and its quality. From the module electrical properties, the temperature effect could be realized by observing the variation of the device parameters:

$$P_{mpp} = I_{sc}V_{oc}FF \tag{22}$$

In the case of Isc and FF, there is very little change that occurred with temperature for crystalline silicon and thin-film devices. Alternatively, the Voc is highly dependent on the temperature variation, which can be described via the Voc as calculated from the one diode model as shown below:

$$\frac{d}{dt}V_{oc} = \frac{d}{dt}\left(n\ V_T\ ln\ \frac{I_{sc}}{I_o}\right) \tag{23}$$

And,

$$V_T = \frac{kT}{q} \tag{24}$$

Where V_T is known as thermal voltage, T is the solar cell temperature, k is the Boltzmann constant, q is the elemental charge q, n is the ideality factor and I_0 is the diode saturation current. From the above diode equation, it could seem that the Voc is positively changing with the temperature, because the above-simplified diode equation typically overlooked the parasitic factors, such as solar cell series and shunt resistance. Particularly, this parasitic resistance is changed significantly over thermal variation [43] and greatly impacts the voltage and diode saturation current as reported elsewhere [44]. For understanding the impact of temperature on Voc, we have to consider the temperature-dependent diode saturation current, which in turn:

$$I_0 = B\ T^\gamma\ exp\left(\frac{E_g}{kT}\right) \tag{25}$$

Where B is a temperature-independent empirical factor but controlled by the quality of absorber material, γ is also an empirical factor that relies on the specific carrier loss mechanism and E_g is the absorber material bandgap. The influence of irradiance and module temperature can be explored by combining the data according to these dependencies. The resulting matrix can then be used to model the annual yield for various technologies at different locations [45]. The main uncertainties, in this case, are kWp standardization and input irradiance [46].

3.4 Solar spectrum distribution model

As there are several uncertainty factors are involving, for the easy and efficient deployment of PV solar cell system, it is essential to measure and develop a model for the spectral distribution of solar radiation. Colle et al. [47] have shown that there has a linear relationship between the uncertainty of solar irradiation and the uncertainty of solar thermal and PV systems. This is a big challenge in the 21st century to develop a more efficient and robust model that could reduce the solar radiation misprint include will need fewer input parameters, will have smaller residual and can be used in a wide variety of conditions.

Indeed, the solar spectrum depends on the place, time, and condition of the atmosphere. The global solar spectrum may be divided into two spectrum models, one for direct beam radiation and the other for diffuse radiation. Particularly, the spectrum of solar incident radiation wavelengths on the PV modules corresponds to the appropriate spectral response range of the PV cells. Several reports on the effect of spectral irradiance variation and PV solar cell performance can be found elsewhere [48, 49]. The longer irradiation hours provided the better annual average electricity outputs [50]. The effect of solar spectral irradiation on the yield of several PV technologies has been documented by Nann and Emery at four separate locations [51]. Eke et al., on the other hand, found that the spectrum variance had a very limited effect on the low bandgap absorber content in PV solar cells [52]. **Figures 6** and **7** shows the spectral response characteristics and EQE of different PV technologies which indicate that how the performance of PV module could change upon the variation spectral distribution.

Several solar spectrum models, including SPECTRAL2 [53], LOWTRAN2 [54], REST2 [55], and SMARTS2 [56], have been developed yet to date over time for clear skies. These models are usually computer programs developed to evaluate the shortwave spectrum components of surface solar irradiance in the range of 280 to 4000 nm. Some of them have high spectral resolutions, however, they need very complex calculations making them less efficient. In the case of LOWTRAN(2), detailed inputs are needed, which increases the execution time and creates some performance limitations, that's why the use of this model is limited in engineering applications [57]. On the other hand, even a low number of parameters are needed for SPECTRAL2, however, the mean deviation associated with different aerosol models is higher than SMART2 [58]. On the other hand, transmittance parameterizations based on the SMARTS spectral model are used to build the high-performance REST2 model [57]. Particularly, more updated parametric functions and constants are used in the SMARTS2 model, for which it has a higher resolution and is showing lower deviation in the spectral analysis. SMARTS program is written in FORTRAN and depends on simplifications of the radiative transfer equation which allow very quick calculations of the irradiance of the surface. The newest versions, such as SMARTS2.9.2 and SMARTS 2.9.5 are hosted by NREL.

The SMART model uses different inputs to define the conditions of the atmosphere under which the irradiance spectra are to be measured. Ideal conditions can also be selected by the user, based on various potential model atmospheres and aerosol models. Moreover, it is also possible to determine practical conditions as inputs, based, for example, on aerosol and water vapor data supplied by a sun photometer [59]. Besides, the spectrally integrated (or 'broadband') irradiance values are given by this model, which can later be compared with measurements from a pyranometer (for diffuse or global radiation) or pyrheliometer (for direct radiation). Solar geometry is another vital input in this model in addition to the atmospheric condition, which is typically specified by the position of the sun (zenith angle and azimuth), the location, the air mass (AM), or by specific time and date. More details on the usage of the

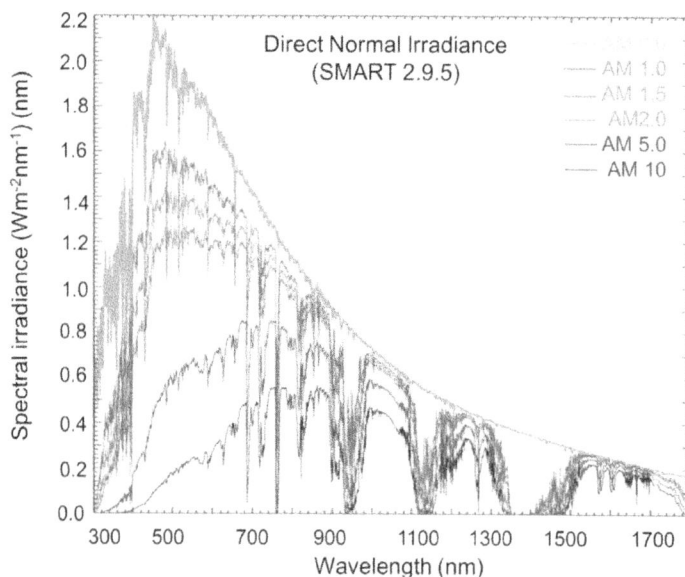

Figure 8.
Direct normal irradiance spectra calculated with SMARTS 2.9.5 for increasing air mass (0 to 10), using the same atmospheric conditions as the ASTM G173 standard. Air mass 0 corresponds to the extraterrestrial spectrum, marked as top of atmosphere (TOA), modified from [67].

SMARTS model for PV applications can be found elsewhere [60–63]. Particularly, this model is frequently employed to evaluate PV modules' efficiency and mismatch factors in real-world conditions [64–66]. **Figure 8** shows the direct normal irradiance spectra with SMART 2.9.5 for different air mass.

4. Performance of PV modules by technologies

Crystalline silicon (c-Si) is the most prevalent PV technology on the market (c-Si). In considering crystal size and crystallinity, c-Si can be divided into two major categories, mono or single-crystalline Si (sc-S) and multi or polycrystalline Si (mc-Si). The power conversion efficiency of sc-Si is higher than mc-Si solar cells, alternatively, sc-Si is costly than mc-Si. The typical efficiency of commercial c-Si modules is between 11% and 20% which power generation varies by temperature (temperature coefficients) in the range of 0.3–0.5%/K [68]. Commercial c-Si modules consisting of 200–500 μm thick PV cells that are connected in series and/or parallel for attaining expected voltage and current. It is important to note that c-Si solar cells or PV modules can generate electrical energy for a wide range of the spectrum (350–1200 nm) as illustrated in **Figure 9** [69]. However, the absorption coefficient of c-Si is below 10^4 cm^{-1} for all wavelengths larger than 500 nm as shown in **Figure 9**. This means that all the potential photons below 500 nm are absorbed close to the surface of the cell. Thus, it is important for the c-Si solar cell that the active region has to be located near the cell surface for absorbing all potential photons and achieving optimum efficiency. Also, it could be seen in **Figure 9** that the absorption coefficient is below 2.0×10^4 for wavelength above 650 nm. As the absorption coefficient of c-Si is below 10^3 for wavelengths above 700 nm which indicates that photons in this range can penetrate the bulk and generate electron–hole pairs. However, their contribution to the photocurrent is very hard in the case of conventional c-Si solar cells. Thus, for collecting these bulk carriers, the configuration of conventional c-Si structure modified, by names they

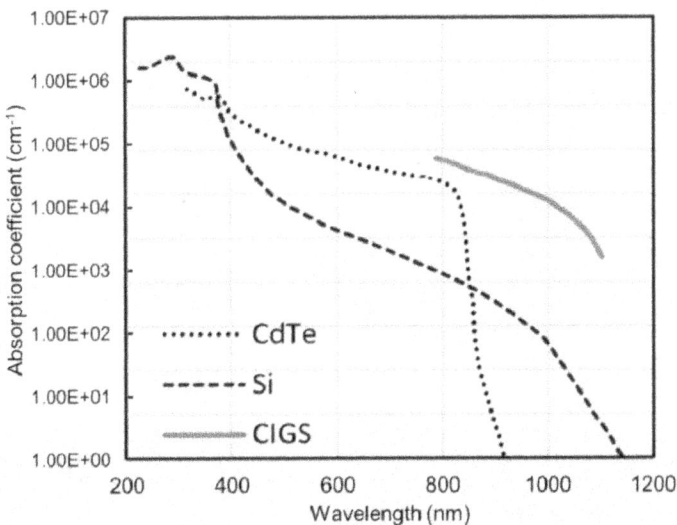

Figure 9.
Absorption spectrum of Si, CdTe, and CIGS solar cells, modified from [69].

are passivated emitter and rear contact (PERC) [70], passivated emitter and rear locally-diffused PERL [71], interdigitated back-contact (IBC) c-Si [72] solar cells.

As it has been mentioned earlier that the response to spectral variation by different types of PV modules vastly depends on its material properties and structure, c-Si solar cells also showed different characteristics depending on the irradiation properties. Several studies have been reported on the in-field energy output analysis of c-Si PV systems by Panchula et al. [73] based in Ontario, Canada; Dolara et al. [74] based in Tuscania, Italy; Fiances et al. [75] Based on a different place in Peru, Kazem et al. [76] based on the desert area of Sohar, Oman, Fuentes et al. [77] and Muñoz et al. [78] under warm climate of Spain, Bahaidarah et al. [79] based on Dhahran, Saudi Arabia and Edalati et al. [80] based on Kerman, Iran. In the above reports, they typically estimate the performance of the system based only on average monthly or yearly insolation and performance ratio varied by the location ranging from around 0.7 to 0.85. Fiances et al. [75] studied different Si technology includes sc-Si, mc-Si, a-Si, and μc-Si PV modules in the climate of Peru, and finalize that a-Si/μc-Si PV modules perform much better than others with an annual performance ratio of 0.97. Ahmed Ghitas [81] reported the effects of the spectral variations on the mc-Si module performance based on outdoor measurements in daily irradiation changes. They only consider cloud-free days in Helwan, Egypt in their measurements and also did not consider the temperature effect. The variation of Voc, Jsc, and power concerning radiation intensity is shown in **Figure 10**. It is evident from **Figure 10** that the most affected device parameter is Isc, and output power in the case of the mc-Si PV module.

Eke and Demircan [82] have been studied mc-Si PV module performance based on winter (January) and Summer (August) for Mugla, Turkey. The operating temperature at this location is 50.5 °C in January and 80.5 °C on August 16. The power generation of the module is 30% lesser in summer than winter because of the significant difference in operating temperature. The power generation every day in January and August is shown in **Figure 11**. Bora et al. [83] also studied the pc-Si PV module along with a-Si, HIT-Si PV modules under the climate condition of the different parts of India. They find that all these three types of Si-based PV modules produce the highest energy yield in the cold and sunny zone.

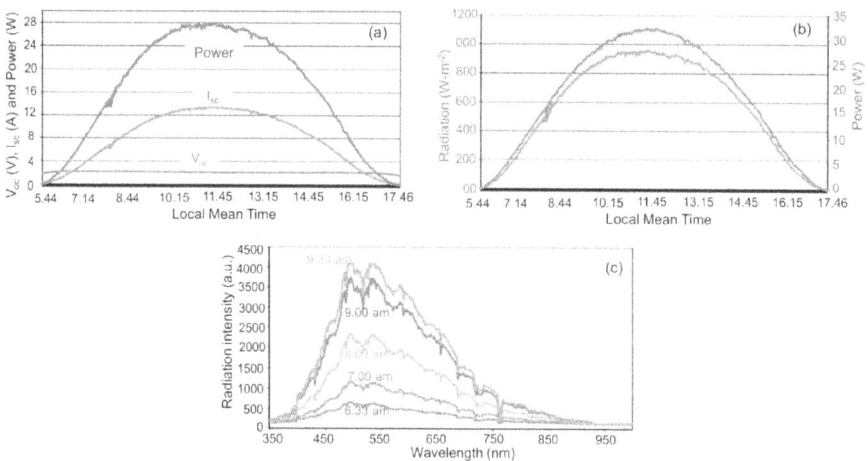

Figure 10.
(a) Daily profile of the measured solar module short circuit current, open-circuit voltage, and electrical output power, (b) daily profile of incident solar radiation along with module output power, and (c) spectral irradiance variation versus time (a.ms) on a clear sky measurement day [81].

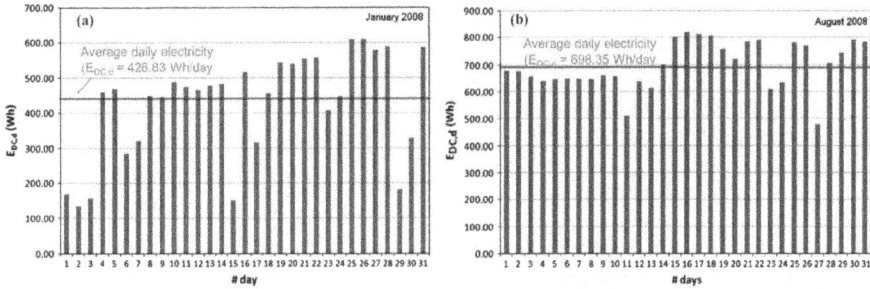

Figure 11.
PV module performance in January (a) and august (b) 2008 for Mugla, Turkey environment [82].

It is important to mention that the energy yield analysis of a PV system is incomplete if their low light condition analysis is missing. Reich et al. [84] have reported the performance of c-Si at low light conditions, however, the impact of temperature is missing as they conducted the study focusing on indoor performance. The finalized that the obtained efficiency via indoor measurement and rated efficiency has a significant difference. Certainly, temperature is a dominant factor in the performance of the PV system in outdoor conditions. It should be noted that solar irradiance and ambient temperature are proportional. Chander et al. [85], and Atsu and Dhaundiyal [86] studied output yield using a detailed model that includes temperature and wind speed variation. Chander et al. [85] reported that the performance parameters of the sc-Si module such as Voc, Pmax, FF, and efficiency are decreased with temperature while the Jsc is increased. Bahaidarah et al. [73] also suggested that for achieving the highest PV performance yield in Saudi Arabia, a suitable and uniform cooling system is necessary due to the climatic conditions. A detailed study on performance variation by low light conditions along with the temperature variation effect has been presented by Pervaiz and Khan [87]. In their

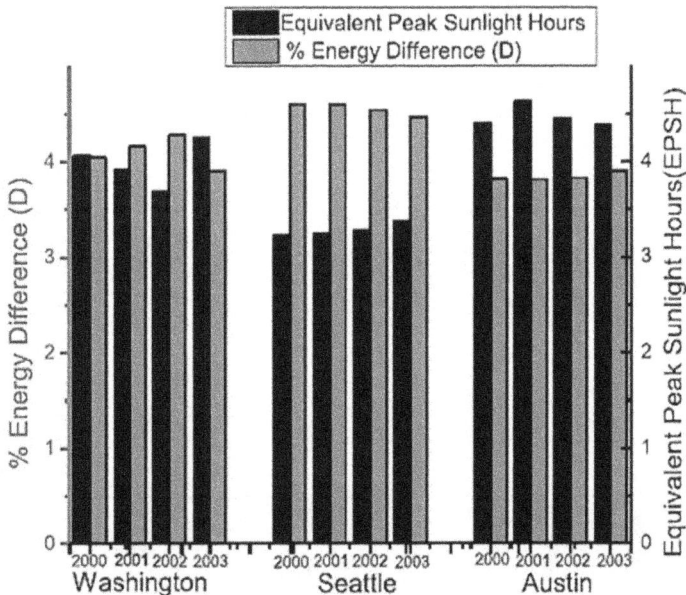

Figure 12.
Energy difference (D) in percent for Washington, Seattle, and Austin for years 2000–2003 [78].

modeling, they used various insulation profiles for a different location in the US collected from NREL. They reported that the energy harvesting of a PV system for a specific location depends on the average peak sunlight hours of that location as shown in **Figure 12**. The use of the following equation for calculating energy difference concerning the variation of Equivalent Peak Sunlight Hours (EPSH).

$$\text{Energy difference, } D = [(Ei - Ec)/Ei] \times 100 \qquad (26)$$

Where Ei is the energy harvested during one year considering a constant efficiency and Ec is the energy harvested incorporating a change in efficiency. The finalized that the reduction in energy yield is reliant on the EPSH of a region where reduction factor could range from 1.5 to 5% for various regions concerning the value of the EPSH.

Cotfas and Cotfas [88] have been studied details on the performance of sc-Si and a-Si PV modules under the natural condition via years of observation, in Brasov, Romania. They reported that the average Pmax of the sc-Si module is two times greater than the a-Si module, however, on clear winter days, the values even increase near to three times greater. Also, at low irradiance, under 100 W/m², the power gain is of sc-Si is 1.9 times greater than a-Si. The gain is over 1.9 times even for very low irradiance, under 100 W/m². The detailed performance of the a-Si PV module including other thin-film modules as an impact of irradiance and temperature are covered in the next section. Under the Mediterranean climatic conditions of the north of Athens, Greece [89], the performance of the p-Si photovoltaic system has been investigated. There is a linear relationship between the module surface temperature and the irradiance where the average temperature about 49.9 °C in summer and 16.8 °C in winter. The efficiency of the p-Si module has been significantly dropped in summer where it ranging from 6.2% to 10.4% concerning the module temperature.

The SR of PV cells depends on the absorption coefficient and/or bandgap of the absorber materials. Similarly, the performance variation by increase or decrease of temperature also depends on the bandgap [90]. The semiconductor material with a wider bandgap, such as 1.04–1.68 eV for CIGS [91], 1.45–1.5 eV for CdTe [92, 93], and (1.7–1.9 eV for a-Si [94] shows higher temperature resistance to the increase of

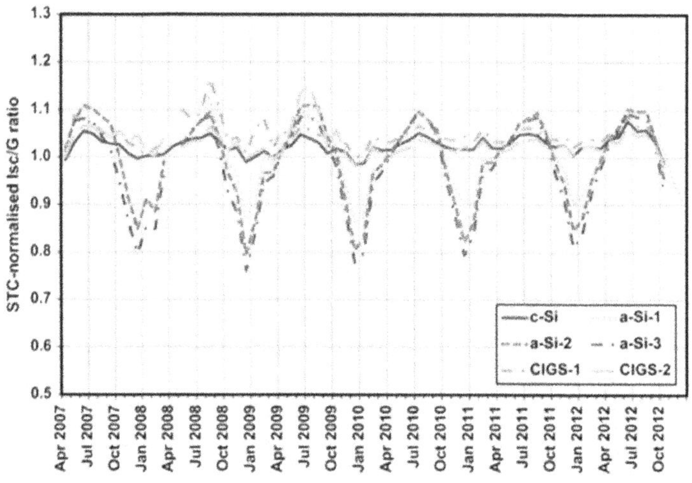

Figure 13.
Calculated spectral effects for the devices under test in the UK environment. The graph compares the normalized ISC divided by the irradiance measured with the pyrometer [86].

module temperature. As a consequence, they have a lower temperature coefficient than sc-Si and pc-Si PV modules [95], and thus, device performance is significantly affected by the temperature. The details on the effects of irradiance, spectrum, and temperature on thin-film PV modules were investigated by Gottschalk et al. [96] under the UK environment. It has been reported that the performance of a-Si is highly spectral dependent as shown in **Figure 13**. The relative change in short circuit current (Isc) is +10% to −20% observed for a-Si whereas the change is only ±3% for c-Si and CIGS. Environmental effects have also been shown to cause up to 15% of losses to the annual PV production. The spectral impact on different PV technologies for all single months has been investigated under the German climate condition [32]. Similar to the other reports, the spectral impact changes more for bigger bandgap a-Si PV modules as shown in **Table 1**. The average gains over the year are 3.4% for a-Si, 1.1% for c-Si, 0.6% for CIGS, and 2.4% for CdTe. It has been reported that CIGS and c-Si modules exhibit high gains in winter and a-Si and CdTe shows an advantage in summer attributed mostly to spectrum variation [32]. The study carried in the Netherlands [97] showed that low irradiance caused a decrease in annual energy yield of 1.2% for the CIGS modules and 1% for CdTe. This experimental study also indicated a strong effect of spectral variation on the performance of the a-Si modules.

The detail on performance variation by the influence of temperature of the different types of PV modules has been conducted by Gutkowski et al. under the low insolation climate of Poland [95]. They observed a significant difference in performance by different PV modules at temperatures 15-48 °C as shown in **Figure 14(a)**. It is clear from **Figure 14(a)** that under real conditions of the high-temperature region, the power generated by CIGS thin-film technologies is higher compare to the pc-Si PV modules. Ozden et al. [98] also experimentally investigate the a-Si and CdTe thin-film PV module performance under the Turkey climate zone along with sc-Si and mc-Si. They found a significant difference in performance in that module for the sunny and cloudy days as shown in **Figure 14(b)**. The output performance of sc-Si and mc-Si is found to be the same, but the output difference

i	Gi [kWh] average monthly irradiation from the reference period	Average, relative monthly spectral impact				
		a-Si (%)	CdTe (%)	c-Si (%)	High-eff. c-Si (%)	CIGS (%)
1	38	−2.0	1	1.9	2.4	2.6
2	65	−1.3	0.1	1	1.4	1.6
3	122	0.1	0.6	0.7	0.8	0.9
4	141	3.5	1.9	1.2	0.9	0.4
5	166	4.2	2.3	1.5	0.9	0.3
6	166	5.1	2.8	1.4	0.8	0
7	184	5.3	3.4	1.5	0.8	0
8	168	5.3	3.5	1.6	0.9	0.1
9	136	4.3	3.1	1.5	1	0.4
10	91	2.8	3	1.9	1.7	1.3
11	43	0.8	2.3	2.1	2.2	2.1
12	35	−2.2	1.8	2.4	3	3.3

Table 1.
Calculation of annual spectral impact based on the monthly sums of irradiance of a reference year and the determined average monthly spectral impact assessed in Germany [32].

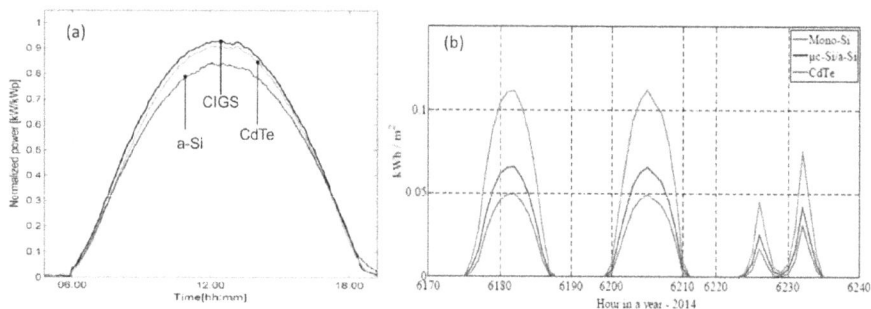

Figure 14.
(a) Normalized DC power generated by the PV systems of each studied technology [95], and (b) maximum irradiance and temperature recorded for that day were 1000 W/m² and 55 °C [78].

between CdTe and sc-Si modules is 60% for a sunny day and which reduces to 35% for a cloudy day indicating the impact of irradiance as well as temperature on these technologies. Moreover, the performance ratio (PR) of sc-Si is in the range of 70%–90%, a-Si is about 70% and CdTe is only 42%–72%. Alternatively, Kesler et al. [99] also conducted a performance analysis between the c-Si and thin film for another location, Antalya, Turkey, and reported that performance of the both technology is very close to each other. Even they specified the reason is the high ambient temperature of that area, however, the rated efficiency of that technologies may play an important role in this case, which means that if the rated efficiency is almost the same, their performance will be close to each other.

Sharma et al. [100] studied three different PV technologies, such as a-Si, pc-Si, and HIT under the tropical climate of India. They found that the best-suited PV technology for this climate is HIT and a-Si. The overall performance ratio for a-Si is 90% and for pc-Si is 83% in this region as shown in **Figure 15**. Interestingly, the energy yield of a-Si is 14% greater during summer, but 6% lower in winter. The effect of seasonal which in turn the effect of irradiance and temperature on the performance of a-Si may be related to its thermal annealing process [101]. The HIT modules have consistently performed better ($\geq$ 4–12%) than p-Si over the year.

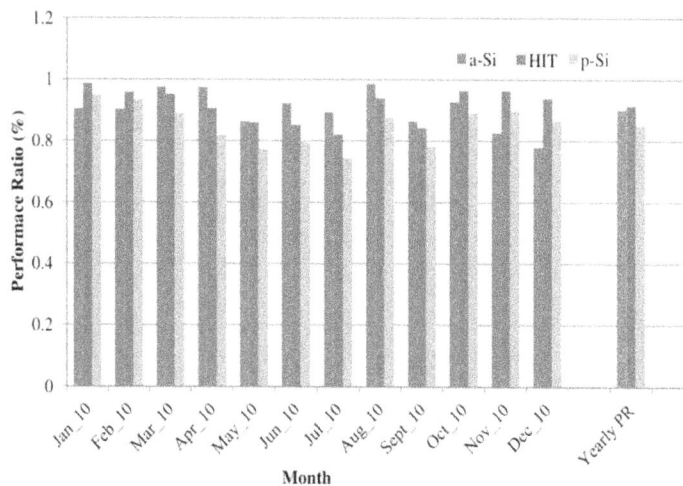

Figure 15.
Comparison of measured monthly and yearly performance ratio of each technology array tested in Indian climate condition [100].

Aste et al. [102] investigate PV module performance under temperate climatic conditions (Italy) where the more distinct seasonal change and/or wide temperature variation have occurred. They found that a-Si is much more sensitive to the seasonal solar spectrum rather than c-Si and HIT technology. The highest 93% of performance ratio has been reported for c-Si in this study. However, the c-Si technology has also shown seasonal variation as an exceptional case [75] and the performance ratio found 20% lower in summer than in winter as a role of temperature variation. In summer, the a-Si/mc-Si stack cell showed higher performance than the other technologies tested in this study, which may be due to its low-temperature coefficient and thermal annealing.

The assessment carried in the Netherlands [103] showed that the CIGS modules are strongly affected by irradiance and temperature variations with a decrease in annual energy yield of 1.2%. Moreover, CdTe modules also exhibited a decrease in energy yield of about 1.0%. This experimental study also showed a significant influence of spectral variation on the efficiency of a-Si modules. Zdyb and Gutkowski studied four different types of PV modules, such as pc-Si, a-Si, CIGS, and CdTe at high latitude under East Poland climate conditions [104]. In their study, a-Si and CIGS shows the gain in performance ratio (about to 73.4% for a-Si and 90.7% for CIGS) during summer sunny and warm environment. The increase of performance ratio of a-Si PV modules has also been reported by Makrides et al. [101] studied under the Cyprus environment. On the other hand, the performance ratio of pc-Si PV modules exhibited over 80% except for December and always remain the highest among the investigated PV modules over the year as shown in **Figure 16**.

The effect of spectral irradiance distribution on the performance of a-Si/mc-Si stacked photovoltaic modules has been analyzed by Minemoto et al. [105] installed at Kusatsu-city (Japan). Their study revealed that these stacked PV modules are extremely spectrally sensitive compared to pc-Si PV modules installed on the same site. Akhmad et al. [106] have been compared the performance of poly-silicon (pc-Si) and amorphous silicon (a-Si) at Kobe, Japan, and found a-Si modules are better for this region. K. Nishioka et al. [107] compared sc-Si, pc-Si module, and heterojunction silicon at Nara Institute of Science and Technology (NAIST) under Japanese climate. They reported that the HIT technology is better suited for this region due to its low-temperature dependency. Poissant [108] has evaluated four

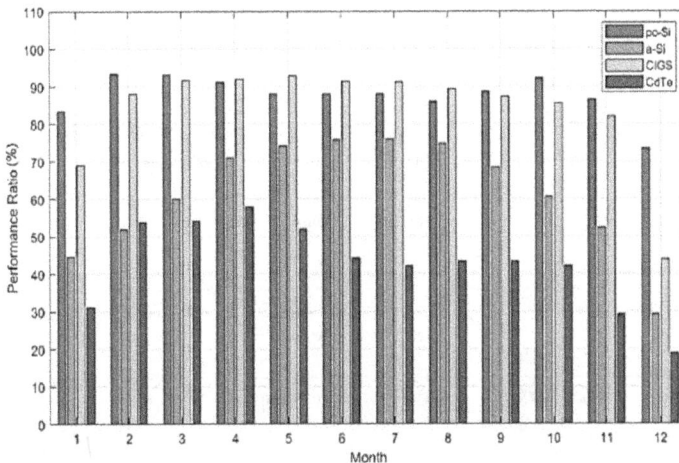

Figure 16.
Performance ratio for each studied PV technology investigated in East Poland (data collected in 2018) [104].

different novels PV module technologies, (i) H-Si, (ii) IBC, (iii) a-Si/uc-Si, and (iv) c-Si under the climate of Canada. His study confirmed that the heterojunction silicon and a-Si/uc-Si technologies are less affected by temperature than the other two crystalline silicon technologies. Canete et al. [109] also performed a comparative study of four different photovoltaic module technologies, (i) amorphous silicon (a-Si), (ii) tandem structure of amorphous silicon- microcrystalline silicon (a-Si/-mc-Si), (iii) polycrystalline silicon module (pc-Si) and (iv) cadmium telluride (CdTe). Their results show that the performance of thin-film modules is better than that of pc-Si modules for the location of Southern Spain. The performances of c-Si

Author(s)	Location	Environmental Parameters	Tested Technologies	Best Perform Technology
Dirnberger et al., 2015 [32]	Breisgau, Germany	Maritime climate, 5-25 °C, 1,117 kW/m²/year (approx.)	a- Si, sc-Si, CIGS and CdTe	a-Si
Francis et al., 2019 [75]	i. Arequipa, Peru Tacna, Peru Lima, Peru	i. Diverse climates 2380 kW/ m2, 3.81-32 °C 2280 kW/ m2, 13.4–31.5 °C 1740 kW/ m2, 18.8–18.9 °C	i. sc-Si ps-Si a- Si/uc-Si	a-Si/μc-Si
Edalati et al., 2015 [81]	Kerman, Iran	Dry climate 68.64– 198.72 kW/m², 20 °C	sc-Si, and pc-Si	pc-Si
Bora et al., 2018 [83]	Different parts of India	0.82–0.87 kW/m²/day not mentioned	a-Si, HIT, and pc- Si	All (cold and sunny zone)
Cotfas and Cotfas, 2019 [88]	Brasov, Romania	Temperate-continental climate, 2.1–1.82 Wh/m2/ day, −4.0 – 24 °C	sc-Si and a-Si	sc-Si
Louwen et al., 2016 [89]	Utrecht, Netherlands	Oceanic climate, 20.5–29.5 ° C, 950–1050 W/m²	SHJ, a- Si, sc-Si, pc-Si, CIGS, CIS and CdTe	sc-Si and SHJ
Gulkowski et al., 2019 [95]	Lublin, Poland	Temperate climate, 950–1250 (kWh/m²)/year, 15–48 °C	CdTe, CIGS, and pc-Si	CIGS
Aste et al., 2014 [94]	Milan, Italy	Temperate climatic, 1270 kW/m²/year, −5-32 °C	c-Si, a-Si/uc-Si, HIT	HIT
Zdyb and Gulkowski, 2020 [103]	Lublin, Poland	Temperate climate, 950–1250 (kWh/m²)/year, 15–48 °C	pc-Si, a-Si, CIGS, and CdTe	pc-Si and CIGS
Makrides et al., 2018 [104]	Cyprus	Mediterranean climate, 1988– 2054 kWh/m², 10-40 °C	sc-Si, pc-Si, a-Si, CIGS and CdTe	a-Si
Minemoto et al., 2007 [105]	Kusatsu-city, Japan	Subtropical climate, 200 kW/ m², 9-33 °C	pc-Si, and a-Si	pc-Si
Poissant, 2009 [108]	Montreal, Canada	Continental climate, 950– 1050 W/m², max. 20 to 22 °C	SHJ, IBC, a- Si/uc-Si, and c-Si	a-Si/uc-Si
Cañete et al., 2014 [109]	Southern Spain	Dry Mediterranean climate, 3.7–7.4 kWh/m²/day, 15–30 °C	a-Si, a-Si/μc-Si CdTe, and pc-Si	a-Si and CdTe

Table 2.
Summary of few reported works for finding out the best PV technology by location and climate.

and a-Si PV modules under South Africa climate conditions have been evaluated by Maluta and Sankaran [110]. They found that both technologies give a similar and suitable performance for the climate of this region. Three different PV technologies (monocrystalline, polycrystalline, and amorphous silicon) have been evaluated under the desert climate by M. Shaltout et al. [111]. They reported that the poly-crystalline silicon cells are more suitable in such a climate. All these above-mentioned studies indicate the difficulty when it comes to choosing the appropriate PV technology for a given site. Thus, the prediction of PV energy potentials before installation helps us to understand the economic advantages associated with it and for policy regulation for electric utilities.

Table 2 shows the summary of a few reported works for finding out the best PV technology by location and its climate. It should be noticed that the results reported by the various researcher as mentioned above are very difficult to compare because the work has been conducted focusing on different locations and various time scales (instantaneous, monthly, annual), different energy effects, and even the works are different by used metrics and calculation. However, it is well agreed that the impact of spectral irradiance variations on PV device performance mostly depends on its spectral response, which in turn depends on its absorber material properties and quality. Moreover, the influence of spectral irradiance on PV performance is dependent on installation sites, for instance, the spectral distribution, climate, environment, latitude, longitude, albedo, etc. of the location. Besides, the spectral distribution of specific sites again depends on the cloudiness, water-vapor and aerosol content in the sky of that sites. The analysis considering all the above factors certainly will be too difficult, thus, the researcher considers only some of the factors for simplifying their work.

5. Conclusion

The weather and/or solar irradiance of the earth is significantly different from one location to another. Again, solar irradiance varies for a specific location by season and/or common weather phenomenon, such as dust, rain, wind, cloud, fog, and snow, etc. Thus, every year solar irradiance also not the same in amount and as an impact of the above factors, the energy yield of different PV technology is affected differently and prediction is very complicated. However, numerous studies could help us to predict which PV technology is better suited for a certain location. It should be noticed that all the incident solar radiations absorbed by PV cells are not able to convert into electricity, some of them are increase temperature, thus the performance varied. As discussed in this book chapter, most of the study showed that summer months when irradiation becomes high that leads to an increase of module temperature, a-Si technology show better performance than c-Si PV mod-ules. It may be due to the metastable defects generated during the dangling bond compensation are decreased upon module temperature increase and as a result, the module performs better in elevated temperature. Also, CIGS PV modules show similar behavior to the a-Si PV modules. The performance gain observed in CIGS technology in summer or at elevated temperatures may be related to the larger bandgap and lower temperature coefficient. Particularly, the optical bandgap of CIGS thin film is higher than a-Si and the higher bandgap has a lower temperature coefficient. Also, CIGS modules can convert the blue light part of the solar spectrum due to a larger bandgap that may assist to perform better in hot summer. Alterna-tively, c-Si have a narrow bandgap, as the defect density increases upon high irradiance and high temperature in hot summer, the dark saturation current and/or leakage current is increased. Consequently, the performance decrease in summer.

However, it has been seen that c-Si perform very in high irradiance with cold weather. It should be noticed that the module with a higher leakage current is highly affected by low irradiance. Since a-Si solar cells inherently have high defect density and/or high leakage current than c-Si solar cells, thus the power gain by c-Si at very low irradiance is significantly higher than a-Si as discussed in the above section. Overall, CdTe modules are performed much poorer than others probably due to the consequence of early degradation of the module as reported in the previous section. All these above-mentioned studies specify the difficulty of choosing an appropriate PV technology for a given site. Thus, the prediction of PV energy potentials before installation is very important concerning the economic advantages and for policy regulation for electric utilities.

Acknowledgements

The authors wish to thank the Ministry of Higher Education of Malaysia (MoHE) for providing the Long Term Research Grant Scheme (LRGS) with the code of LRGS/1/2019/UKM-UNITEN/6/2 to support this research. The authors also acknowledge the publication support from the iRMC of Universiti Tenaga Nasional (@UNITEN), Malaysia. The authors would also like to acknowledge the Faculty of Engineering, University of Malaya (@UM) for other supports.

Author details

Mohammad Aminul Islam[1,2*], Nabilah M. Kassim[2,3], Ammar Ahmed Alkahtani[2,3] and Nowshad Amin[2,3]

1 Department of Electrical Engineering, Faculty of Engineering, University of Malaya, Jalan Universiti, Kuala Lumpur, Selangor, Malaysia

2 Institute of Sustainable Energy, Universiti Tenaga Nasional (@The Energy University), Jalan IKRAM-UNITEN, Kajang, Selangor, Malaysia

3 College of Engineering, Universiti Tenaga Nasional (@The Energy University), Jalan IKRAM-UNITEN, Kajang, Selangor, Malaysia

*Address all correspondence to: aminul.islam@um.edu.my

IntechOpen

References

[1] Ye, J.Y.; Reindl, T.; Aberle, A.G.; Walsh, T.M. Effect of solar spectrum on the performance of various thin-film PV module technologies in tropical Singapore. IEEE J. Photovolt. 2014, 4, 1268–1274.

[2] Alonso-Abella, M.; Chenlo, F.; Nofuentes, G.; Torres-Ramirez, M. Analysis of spectral effects on the energy yield of different PV (photovoltaic) technologies: The case of four specific sites. Energy 2014, 67, 435–443.

[3] Minemoto, T.; Nagae, S.; Takakura, H. Impact of spectral irradiance distribution and temperature on the outdoor performance of amorphous Si photovoltaic modules. Solar Energy Mater. Solar Cells 2007, 91, 919–923.

[4] Gracia Amillo, A.; Huld, T.; Vourlioti, P.; Müller, R.; Norton, M. Application of satellite-based spectrally resolved solar radiation data to PV performance studies. Energies 2015, 8, 3455–3488.

[5] Huld, T.A.; Friesen, G.; Skoczek, A.; Kenny, R.A.; Sample, T.; Field, M.; Dunlop, E.D. A power-rating model for crystalline silicon PV modules. Solar Energy Mater. Solar Cells 2011, 95, 3359–3369.

[6] Huld, T.; Gottschalg, R.; Beyer, H.; Topiˇc, M. Mapping the performance of PV modules, effects of module type and data averaging. Solar Energy 2010, 84, 324–338.

[7] Koehl, M.; Heck, M.; Wiesmeier, S.; Wirth, J. Modeling of the nominal operating cell temperature based on outdoor weathering. Solar Energy Mater. Solar Cells 2011, 95, 1638–1646.

[8] IEC Central Office. Photovoltaic Devices—Part 3: Measurement Principles for Terrestrial Photovoltaic (PV) Solar Devices with Reference Spectral Irradiance Data; Technical Report IEC 61215–3; International Electrotechnical Commission: Geneva, Switzerland, 2005.

[9] World Bank. 2017. Global Solar Atlas. https://globalsolaratlas.info.

[10] "RReDC Glossary of Solar Radiation Resource Terms". Rredc.nrel.gov. Retrieved 25 November 2017.

[11] Kalogirou, Soteris, ed. McEvoy's handbook of photovoltaics: fundamentals and applications. Academic Press, 2017.

[12] International Electro-Technical Commission. Standard IEC 60904–3: Photovoltaic Devices. Part 3: Measurement Principles for Terrestrial Photovoltaic (PV) Solar Devices With Reference Spectral Irradiance Data (Ed. 2, 2008)

[13] Wikimedia Commons. (April 16, 2018). Solar Spectrum [Online]. Available: https://commons.wikimedia.org/wiki/File:Solar_Spectrum.png

[14] Duffie JA, Beckman WA. Solar engineering of thermal processes. John Wiley&Sons Inc., Hoboken, New Jersey, 2006.

[15] Hassanzadeh BH, de Keizer AC, Reich NH, van Sark WGJHM. The effect of a varying solar spectrum on the energy performance of solar cells. In: proceedings of 22nd European PVSEC, Milan, 2007. p. 2652–2658.

[16] Ma, J., Man, K. L., Ting, T. O., Zhang, N., Guan, S. U., & Wong, P. W. (2013). Approximate single-diode photovoltaic model for efficient IV characteristics estimation. The Scientific World Journal, 2013.

[17] Wu, N., Wu, Y., Walter, D., Shen, H., Duong, T., Grant, D., ... & Catchpole, K. (2017). Identifying the

cause of voltage and fill factor losses in perovskite solar cells by using luminescence measurements. Energy Technology, 5(10), 1827–1835.

[18] Luque A, Hegedus S. Handbook of photovoltaic science and engineering. John Wiley & Sons Inc., England, 2003.

[19] Gottschalg R, Betts TR, Infield DG, Kearney MJ, The effect of spectral variations on the performance parameters of single and double junction amorphous silicon solar cells. Solar Energy Materials & Solar Cells 2005; 85: 415–428.

[20] Shimokawa R, Miyake Y, Nakanishi Y, Kuouano Y, Hamakawa Y. Effect of atmospheric parameters on solar cell performance under global irradiance. Solar Cells1986–1987; 19: 59–72.

[21] Rosell JI, Ibanez M, Modelling power output in photovoltaic modules for outdoor operating conditions. Energy Conversion and Management 2006; 47: 2424–2430.

[22] Sadok M, Mehdaoui A. Outdoor testing of photovoltaic arrays in the saharan region. Renewable Energy 2008; 33: 2516–2524.

[23] Kenny RP, Ioannides A, Müllejans H, Zaaiman W, Dunlop ED. Performance of thin film PV modules. Thin Solid Films 2006; 511–512: 663–672.

[24] Shaltout, M.A.M., El-Nicklawy, M., Hassan, A., Rahoma, U., Sabry, M., 2000. The temperature dependence of the spectral and efficiency behavior of Si solar cell under low concentrated solar radiation. Renew. Energy 21, 445–458.

[25] Silvestre, S., Sentı's, L., Castaner, L., 1999. A fast low-cost solar cell spectral response measurement system with accuracy indicator. IEEE Trans. Instrum. Meas. 48 (5), 944–948.

[26] Makrides G, Zinsser B, Phinikarides A, Schubert M, Georghiou GE. Temperature and thermal annealing effects on different photovoltaic technologies. Renewable Energy 2012; 43: 407–417.

[27] Jiang JA, Wang JC, Kuo KC, Su YL, Shieh JC, Chou JJ. Analysis of the junction temperature and thermal characteristics of photovoltaic modules under various operation conditions. Energy 2012; 44: 292–301.

[28] Kamkird P, Ketjoy N, Rakwichian W, Sukchai S. Investigation on temperature coefficients of three types photovoltaic module technologies under Thailand operating condition. Procedia Engineering 2012; 32: 376–383.

[29] Mekhilef S, Saidur R, Kamalisarvestani M. Effect of dust, humidity and air velocity on efficiency of photovoltaic cells. Renewable and Sustainable Energy Reviews 2012; 16: 2920–2925.

[30] Singh P, Ravindra NM. Temperature dependence of solar cell performance-an analysis. Solar Energy Materials & Solar Cells 2012; 101: 36–45.

[31] Okullo W, Munji MK, Vorster FJ, van Dyk EE. Effects of spectral variation on the device performance of copper indium diselenide and multi-crystalline silicon photovoltaic modules. Solar Energy Materials & Solar Cells 2011; 95: 759–764.

[32] Dirnberger D, Blackburn G, Müller B, Reise C. On the impact of solar spectral irradiance on the yield of different PV technologies. Solar Energy Materials and Solar Cells 2015; 132: 431–442.

[33] Alonso-Abella M, Chenlo F, Nofuentes G, Torres- Ramírez M. Analysis of spectral effects on the energy yield of different PV (photovoltaic) technologies: the case of

four specific sites. Energy 2014; 67: 435–443.

[34] Nofuentes G, García-Domingo B, Muñoz JV, Chenlo F. Analysis of the dependence of the spectral factor of some PV technologies on the solar spectrum distribution. Applied Energy 2014; 113: 302–309.

[35] Huld T, Amillo AMG. Estimating PV module performance over large geographical regions: the role of irradiance, air temperature, wind speed and solar spectrum. Energies 2015; 8(6): 5159–5181.

[36] Ishii T, Otani K, Itagaki A, Utsunomiya K. A simplified methodology for estimating solar spectral influence on photovoltaic energy yield using average photon energy. Energy Science & Engineering 2013; 1(1): 18–26.

[37] Cornaro C, Andreotti A. Influence of average photon energy index on solar irradiance characteristics and outdoor performance of photovoltaic modules. Progress in Photovoltaics: Research and Applications 2013; 21(5): 996–1003.

[38] Betts TR, Jardine CN, Gottschalg R, Infield DG, Lane K. Impact of spectral effects on the electrical parameters of multijunction amorphous silicon cells, In Proceedings of 3rd World Conference on Photovoltaic Energy Conversion, 2003, 2003; 1756–1759.

[39] Ishii T, Otani K, Itagaki A, Utsunomiya K. A methodology for estimating the effect of solar spectrum on photovoltaic module performance by using average photon energy and a water absorption band. Japanese Journal of Applied Physics 2012; 51: 51(10NF05).

[40] Beard, M. C., Johnson, J. C., Luther, J. M., & Nozik, A. J. Multiple exciton generation in quantum dots versus singlet fission in molecular chromophores for solar photon conversion. Philosophical Transactions of the Royal Society A: Mathematical, Physical and Engineering Sciences, 2015; 373(2044): 20140412.

[41] Minnaert, B., & Veelaert, P. A proposal for typical artificial light sources for the characterization of indoor photovoltaic applications. Energies, 2014; 7(3): 1500–1516.

[42] Wild, M.; Folini, D.; Henschel, F.; Fischer, N.; Müller, B. Projections of long-term changes in solar radiation based on CMIP5 climate models and their influence on energy yields of photovoltaic systems. Sol. Energy, 2015; 116: 12–24.

[43] R. Gottschalg, M. Rommel, D. G. Infield, Variation of solar cell equivalent circuit parameters under different operating conditions, in: Proceedings of the 14th European Photovoltaic Solar Energy Conference, WIP-Munich, Barcelona,1997, pp.2176–2179.

[44] Hubin, J., & Shah, A. V. Effect of the recombination function on the collection in a p—i—n solar cell. Philosophical Magazine B, 1995; 72(6): 589–599.

[45] Huld, T., Gottschalg, R., Beyer, H. G., & Topič, M. Mapping the performance of PV modules, effects of module type and data averaging. Solar Energy, 2010; 84(2): 324–338.

[46] Gottschalg, R., Betts, T. R., Eeles, A., Williams, S. R., & Zhu, J. Influences on the energy delivery of thin film photovoltaic modules. Solar energy materials and solar cells, 2013; 119: 169–180.

[47] Colle S, De Abreu, SL, Ruther R. Uncertainty in economic analysis of solar water heating and photovoltaic systems. Solar Energy, 2001; 70(2): 131–142.

[48] Prentice, J.S.C. Spectral response of a-Si:H p-i-n solar cells. Sol. Energy Mater. Sol. Cells 2001; 69: 303–314.

[49] Dirnberger, D.; Blackburn, G.; Muller, B.; Reise, C. On the impact of solar spectral irradiance on the yield of different PV technologies. Sol. Energy Mater. Sol. Cells 2015; 132: 431–442.

[50] Tsao, J., Lewis, N. and Crabtree, G. Solar faqs. US department of Energy, 13, 2006.

[51] Nann, S.; Emery, K. Spectral effects on PV-device rating. Sol. Energy Mater. Sol. Cells 1992; 27: 189–216.

[52] Eke, R.; Betts, T.R.; Gottschalg, R. Spectral irradiance effect on the outdoor performance of photovoltaic modules. Renew. Sustain. Energy Rev. 2017; 69: 429–434.

[53] Bird, R.E. A simple, solar spectral model for direct-normal and di use horizontal irradiance. Sol. Energy 1984; 32: 461–471.

[54] McClatchey, R.A.; Selby, J.E. Atmospheric Transmittance from 0.25 to 28.5 lm: Computer Code LOWTRAN2; AFCRL-72-0745, Environ. Res. Paper No. 427; Airforce Cambridge Research Laboratories: Wright-Patterson Air Force Base, OH, USA, 1972.

[55] Gueymard, C.A. REST2, High-performance solar radiation model for cloudless-sky irradiance, illuminance, and photosynthetically active radiation —Validation with a benchmark dataset. Sol. Energy 2008; 82: 272–285.

[56] Gueymard, C.A. SMARTS2, a Simple Model. of the Atmospheric Radiative Transfer of Sunshine: Algorithms and Performance Assessment; FSEC-PF-270-95; Florida Solar Energy Center: Cocoa, FL, USA, 1995.

[57] Gueymard, C.A., REST2: High-performance solar radiation model for cloudless-sky irradiance, illuminance, and photosynthetically active radiation–

Validation with a benchmark dataset. Solar Energy, 2008; 82(3): 272–285.

[58] Utrillas, M.P., Bosca, J.V., Martínez-Lozano, J.A., Cañada, J., Tena, F. and Pinazo, J.M. A comparative study of SPCTRAL2 and SMARTS2 parameterised models based on spectral irradiance measurements at Valencia, Spain. Solar energy, 1998; 63(3): 161–171.

[59] Gueymard, C.A. Interdisciplinary applications of a versatile spectral solar irradiance model: A review. Energy, 2005; 30(9): 1551–1576.

[60] Myers, D.R., Emery, K. and Gueymard, C. Revising and validating spectral irradiance reference standards for photovoltaic performance evaluation. J. Sol. Energy Eng., 2004; 126(1): 567–574.

[61] Philipps, S.P., Peharz, G., Hoheisel, R., Hornung, T., Al-Abbadi, N.M., Dimroth, F. and Bett, A.W. Energy harvesting efficiency of III–V triple-junction concentrator solar cells under realistic spectral conditions. Solar Energy Materials and Solar Cells, 2010; 94(5): 869–877.

[62] Jaus, J. and Gueymard, C.A. Generalized spectral performance evaluation of multijunction solar cells using a multicore, parallelized version of SMARTS. In AIP Conference Proceedings, 2012; 1477(1): 122–126).

[63] Marion, B. Preliminary investigation of methods for correcting for variations in solar spectrum under clear skies, 2010.

[64] Guechi, A. and Chegaar, M. Effects of diffuse spectral illumination on microcrystalline solar cells. J. Electron Devices, 2007; 5: 116–121.

[65] Dobbin, A., Norton, M., Georghiou, G.E., Lumb, M. and Tibbits, T.N. December. Energy harvest predictions

for a spectrally tuned multiple quantum well device utilising measured and modelled solar spectra. In AIP Conference Proceedings, 2011; 1407(1): 21–24).

[66] Muller, M., Marion, B., Kurtz, S. and Rodriguez, J. An investigation into spectral parameters as they impact CPV module performance. In AIP conference proceedings, 2010; 1277(1): 307–311).

[67] Gueymard, C.A., The sun's total and spectral irradiance for solar energy applications and solar radiation models. Solar energy, 2004; 76(4): 423–453.

[68] Monokroussos, C., Zhang, X. Y., Schweiger, M., Etienne, D., Liu, S., Zhou, A., ... & Zou, C. (2017). Energy Rating of c-Si and mc-Si Commercial PV-Modules in Accordance with IEC 61853–1,-2,-3 and Impact on the Annual Yield. In Proceedings of 33rd European Photovoltaic Solar Energy Conference.

[69] Habibi, M., Zabihi, F., Ahmadian-Yazdi, M. R., & Eslamian, M. (2016). Progress in emerging solution-processed thin film solar cells–Part II: Perovskite solar cells. Renewable and Sustainable Energy Reviews, 62, 1012–1031.

[70] Zhang, C., Shen, H., Sun, L., Yang, J., Wu, S., & Lu, Z. (2020). Bifacial p-Type PERC Solar Cell with Efficiency over 22% Using Laser Doped Selective Emitter. Energies, 13(6), 1388.

[71] Zhao, J., Wang, A., & Green, M. A. (2001). High-efficiency PERL and PERT silicon solar cells on FZ and MCZ substrates. Solar Energy Materials and Solar Cells, 65(1–4), 429–435.

[72] Vasudevan, R., Harrison, S., D'Alonzo, G., Moustafa, A., Nos, O., Muñoz, D., & Roux, C. (2018, August). Laser-induced BSF: A new approach to simplify IBC-SHJ solar cell fabrication. In AIP Conference Proceedings (Vol. 1999, No. 1, p. 040024). AIP Publishing LLC.

[73] A. F. Panchula et al., "First year performance of a 20 MWac PV power plant," in 37th IEEE Photovoltaic Specialists Conference (PVSC) (2011).

[74] A. Dolara et al., "Performance analysis of a single-axis tracking PV system," IEEE J. Photovolt. 2(4), 524–531 (2012).

[75] Romero-Fiances, I., Muñoz-Cerón, E., Espinoza-Paredes, R., Nofuentes, G., & De la Casa, J. (2019). Analysis of the performance of various pv module technologies in Peru. Energies, 12(1), 186.

[76] Kazem, H.A.; Khatib, T.; Sopian, K.; Elmenreich, W. Performance and feasibility assessment of a 1.4 kW roof top grid-connected photovoltaic power system under desertic weather conditions. Energy Build. 2014, 82, 123–129

[77] Fuentes, M.; Nofuentes, G.; Aguilera, J.; Talavera, D.L.; Castro, M. Application and validation of algebraic methods to predict the behaviour of crystalline silicon PV modules in Mediterranean climates. Sol. Energy 2007, 81, 1396–1408.

[78] Muñoz, J.V.; Nofuentes, G.; Fuentes, M.; de la Casa, J.; Aguilera, J. DC energy yield prediction in large monocrystalline and polycrystalline PV plants: Time-domain integration of Osterwald's model. Energy 2016, 114, 951–960.

[79] Bahaidarah, H., Rehman, S., Subhan, A., Gandhidasan, P., & Baig, H. (2015). Performance evaluation of a PV module under climatic conditions of Dhahran, Saudi Arabia. Energy exploration & exploitation, 33(6), 909–929.

[80] Edalati, S.; Ameri, M.; Iranmanesh, M. Comparative performance investigation of mono- and poly-crystalline silicon photovoltaic modules for use in grid-connected photovoltaic systems in dry climates. Appl. Energy 2015, 160, 255–265.

[81] Ghitas, A. E. (2012). Studying the effect of spectral variations intensity of the incident solar radiation on the Si solar cells performance. NRIAG Journal of Astronomy and Geophysics, 1(2), 165–171.

[82] Eke, R.; Demircan, H. Performance analysis of a multi crystalline Si photovoltaic module under Mugla climatic conditions in Turkey. Energy Convers. Manag. 2013, 65, 580–586.

[83] Bora, B.; Kumar, R.; Sastry, O.S.; Prasad, B.; Mondal, S.; Tripathi, A.K. Energy rating estimation of PV module technologies for different climatic conditions. Sol. Energy 2018, 174, 901–911.

[84] N. J. Reich et al., "Crystalline silicon cell performance at low light intensities," Sol. Energy Mater. Sol. Cells 93(9), 1471–1481 (2009)

[85] Chander, S., Purohit, A., Sharma, A., Nehra, S. P., & Dhaka, M. S. (2015). Impact of temperature on performance of series and parallel connected mono-crystalline silicon solar cells. Energy Reports, 1, 175–180.

[86] Atsu, D., & Dhaundiyal, A. (2019). Effect of Ambient Parameters on the Temperature Distribution of Photovoltaic (PV) Modules. Resources, 8(2), 107.

[87] Pervaiz, S., & Khan, H. A. (2015). Low irradiance loss quantification in c-Si panels for photovoltaic systems. Journal of Renewable and Sustainable Energy, 7(1), 013129.

[88] Cotfas, D. T., & Cotfas, P. A. (2019). Comparative Study of Two Commercial Photovoltaic Panels under Natural Sunlight Conditions. International Journal of Photoenergy, 2019.

[89] Gaglia, A.G.; Lykoudis, S.; Argiriou, A.A.; Balaras, C.A.; Dialynas, E. Energy efficiency of PV panels under real outdoor conditions—An experimental assessment in Athens, Greece. Renew. Energy 2017, 101, 236–243

[90] Luceño-Sánchez, J. A., Díez-Pascual, A. M., & Peña Capilla, R. (2019). Materials for photovoltaics: State of art and recent developments. International journal of molecular sciences, 20(4), 976.

[91] Shafarman, W.N.; Stolt, L. Cu (InGa)Se2 solar cells. In Handbook of Photovoltaic Science and Engineering; Luque, A., Hegedus, S., Eds.;Wiley: Hoboken, NJ, USA, 2003; pp. 567–616.

[92] Khana, N.A.; Rahmanb, K.S.; Aris, K.A.; Ali, A.M.; Misran, H.; Akhtaruzzaman, M.; Tiong, S.K.; Amin, N. Effect of laser annealing on thermally evaporated CdTe thin films for photovoltaic absorber application. Sol. Energy 2018, 173, 1051–1057.

[93] Islam, M. A., Rahman, K. S., Sobayel, K., Enam, T., Ali, A. M., Zaman, M., ... & Amin, N. (2017). Fabrication of high efficiency sputtered CdS: O/CdTe thin film solar cells from window/absorber layer growth optimization in magnetron sputtering. Solar Energy Materials and Solar Cells, 172, 384–393.

[94] Morigaki, K.; Ogihara, C. Amorphous Semiconductors: Structure, Optical and Electrical Properties. In Springer Handbook of Electronic and Photonic Materials; Kasap, S., Capper, P., Eds.; Springer: Cham, Germany, 2017.

[95] Gulkowski, S., Zdyb, A., & Dragan, P. (2019). Experimental efficiency analysis of a photovoltaic system with different module technologies under temperate climate conditions. Applied Sciences, 9(1), 141.

[96] Gottschalg, R.; Betts, T.R.; Eeles, A.; Williams, A.R.; Zhu, J. Influences on the energy delivery of thin film photovoltaic modules. Sol. Energy Mater. Sol. Cells 2013, 119, 169–180.

[97] Louwen, A.; de Waal, A.C.; Schropp, R.E.I.; Faaij, A.P.C.; van Sark, W.G.J.H.M. Comprehensive characterization and analysis of PV module performance under real operating conditions. Prog. Photovolt. Res. Appl. 2017, 25, 218–232.

[98] Ozden, T., Akinoglu, B. G., & Turan, R. (2017). Long term outdoor performances of three different on-grid PV arrays in central Anatolia–An extended analysis. Renewable energy, 101, 182–195.

[99] Kesler, S.; Kivrak, S.; Dincer, F.; Rustemli, S.; Karaaslan, M.; Unal, E.; Erdiven, U. The analysis of PV power potential and system installation in Manavgat, Turkey— A case study in winter season. Renew. Sustain. Energy Rev. 2014, 31, 671–680.

[100] Sharma, V.; Kumar, A.; Sastry, O. S.; Chandel, S.S. Performance assessment of different solar photovoltaic technologies under similar outdoor conditions. Energy 2013, 58, 511–518.

[101] Makrides G, Zinsser B, Phinikarides A, Schubert M, Georghiou GE. Temperature and thermal annealing effects on different photovoltaic technologies. Renewable Energy 2012;43:407e17.

[102] Aste, N.; Del Pero, C.; Leonforte, F. PV technologies performance comparison in temperate climates. Sol. Energy 2014, 109, 1–10.

[103] Louwen, A.; de Waal, A.C.; Schropp, R.E.I.; Faaij, A.P.C.; van Sark, W.G.J.H.M. Comprehensive characterization and analysis of PV module performance under real operating conditions. Prog. Photovolt. Res. Appl. 2017; 25: 218–232

[104] Zdyb, A., & Gulkowski, S. (2020). Performance Assessment of Four Different Photovoltaic Technologies in Poland. Energies, 13(1), 196.

[105] Minemoto, T., Toda, M., Nagae, S., Gotoh, M., Nakajima, A., Yamamoto, K., ... & Hamakawa, Y. (2007). Effect of spectral irradiance distribution on the outdoor performance of amorphous Si//thin-film crystalline Si stacked photovoltaic modules. Solar Energy Materials and Solar Cells, 91(2–3), 120–122.

[106] Akhmad K, Kitamura A, Yamamoto F, Okamoto H, Takakura H, Hamakawa Y. Outdoor performance of amorphous silicon and polycrystalline silicon PV modules. Solar Energy Materials and Solar Cells 1997; 46(3): 209–218.

[107] Nishioka, K., Hatayama, T., Uraoka, Y., Fuyuki, T., Hagihara, R., & Watanabe, M. (2003). Field-test analysis of PV system output characteristics focusing on module temperature. Solar Energy Materials and Solar Cells, 75(3–4), 665–671.

[108] Poissant, Y. (2009, June). Field assessment of novel PV module technologies in Canada. In Proc. 4th Canadian Solar Buildings Conference, June.

[109] Cañete, C., Carretero, J., & Sidrach-de-Cardona, M. (2014). Energy performance of different photovoltaic module technologies under outdoor conditions. Energy, 65, 295–302.

[110] Maluta, E., & Sankaran, V. (2011). Outdoor testing of amorphous and crystalline silicon solar panels at Thohoyandou. Journal of Energy in Southern Africa, 22(3), 16–22.

[111] Shaltout, M. M., El-Hadad, A. A., Fadly, M. A., Hassan, A. F., & Mahrous, A. M. (2000). Determination of suitable types of solar cells for optimal outdoor performance in desert climate. Renewable energy, 19(1–2), 71–74.

Chapter 7

Outdoor Performance and Stability Assessment of Dye-Sensitized Solar Cells (DSSCs)

Reema Agarwal, Yogeshwari Vyas, Priyanka Chundawat, Dharmendra and Chetna Ameta

Abstract

In this era the requirement for energy is enhancing, therefore, many energy resources are developed among them the emerging third-generation dye-sensitized solar cell is one of the environment-friendly solar cell-based technology. Generally, dye-sensitized solar cells consist of a nanomaterial-based photoanode, dye molecules as an absorber, electrolyte, and counter electrode. In the case of indoor application, this solar cell works easily so this is the characteristics of a dye-sensitized solar cell. Moreover, the outdoor performance of DSSC degrades on exposure to sunlight. Exposure to sunlight increases the temperature of the internal component of DSSC and consequently degradation in device performance. Long-term stability is obtained by the choice of such material where degradation takes place slowly and plastic covers are also coated over DSSC to prevent degradation. The solar response of DSSC towards dye was also mentioned, the higher the percentage of EQE higher the efficiency of the device. In this chapter, the authors discuss the introduction of a solar cell, the working principle of DSSC, and the available research background for outdoor performance and long-term stability with a solar response of device i.e. EQE or IPCE.

Keywords: DSSC, Solar energy, Outdoor, Stability, IPCE

1. Introduction

Climate change in the 21st century influences the water resources and food which pattern disease and impact greatly the mankind livelihood. Thus, an efficient mechanism is mandatory to control the emission of hazardous gases. The reduction in carbon emission will also help greatly for the environment, most of the nations are seriously working to mitigate this problem. The utilization of available low Carbon energy resources such as solar and wind will be a milestone to cater to the energy necessities of the globe without harming the environment. After the oil crisis in the year 1973, the alternative sources for energy harvesting are derived by many scientists and still, research is going on [1]. The rising population and higher living standards are influencing climate change significantly. Industrialization, the technologically driven changing landscape of cities have increased the energy demand hugely. The resources of energy are commercial and non-commercial where the commercial resources mainly include fossil fuels like coal or natural gas whereas the non-commercial resources include wood and animal and agriculture wastes as well. Fossil fuel resources are non-renewable, limited in stocks, and creates pollution in the environment, as well as these, are fastly depleting. Therefore, research on the development of new energy resources is extremely needed to cater to the energy demand of the revolutionized world. Renewable energy resources are eco-friendly, abundant, and practically inexhaustible. Sun is one of the renewable resources for green and free energy which provides a tremendous amount of energy without any expenditure. The sun irradiates more energy per hour vis-à-vis the total energy consumed globally during one year. Solar energy is non-depletable, pollution-free, and available in abundance on the surface of Earth planet throughout the year. The Bloomberg New Energy Finance (BNEF) research organization made research on the current scenario of energy consumption and production and concluded that 50 percent of the world's energy would come from solar cells and wind by the end of 2050 [2]. Therefore, the use of solar energy could increase the economic growth of any country without affecting the environment.

The solar cell is a device that transforms solar or light into electrical energy, it is just a p-n junction or a diode. The Silicon-based solar cells were firstly used to convert sunlight into electricity, therefore, these solar cells are also recognized as traditional or conventional solar cells. The solar cells are classified into three generations. The first generation or crystalline Silicon solar cells are widely used as these have been shown higher power conversion efficiency (η) about 26% [3, 4] and dominated the solar cell market ever since its invention, but fabrication of crystalline Silicon solar cells suffers from high module cost and a significant amount of by-products. The second generation comprises thin-film-based solar cells which reduced materials consumption and consequently cost of the device. This generation includes amorphous silicon solar cells, cadmium telluride (CdTe) thin-film solar cells, and copper indium gallium diselenide (CIGS) thin-film solar cells [5–7]. The materials to the second generation solar cells are rare elements (e.g. Tellurium) and hazardous (e.g. Cadmium). Due to the high cost of first-generation solar cells, and toxicity, and limited availability of materials for second-generation solar cells, a new generation of solar cells emerges as third generation [8]. The third-generation solar cells comprise a variety of new materials besides the evergreen and champion Silicon which include nanomaterials and Silicon wires. The third-generation solar cells are designed to trim down the cost and are based on the simple, cheap, and easy fabrication process. This generation includes dye-sensitized, polymer, quantum dot, perovskite solar cells. Given cost-effectiveness, efficiency, and easy fabrication process, the dye-sensitized solar cells (DSSC) could be one of the best promising alternatives to the Silicon solar cells [9].

The configuration of dye-sensitized solar cells (DSSCs) comprises a glass substrate (conductive substrate), nanostructure semiconductor (photo-anode), sensitizer (dye), electrolyte, and catalyst counter electrode [10]. Nowadays, the DSSC devices are developed to have such a photo-anode that could efficiently harvest the energy, increase the dye pickup, light scattering ability, reduce recombination reaction and improve charge transferability [11]. The prototype DSSC was reported by Michael Gratzel in 1991. The DSSCs are one of the most efficient photo-to-electron conversion devices under indoor and low-level outdoor lighting for integrating green buildings. For a DSSC device, the highest achieved efficiency is 14.30% (practically) to the date where the Co (II/III) based electrolyte was used with the co-sensitization of organic dyes [12]. The theoretically predicted maximum efficiency for DSSC is 32% which is estimated and limited by the Shockley-Queisser limit based upon the principle of detailed balance [13]. In the architecture of dye-sensitized solar cells, usually, TiO_2 (titanium dioxide) is preferred because of its photoactive, low cost, and abundant availability [14]. The most used dye for DSSC is N719 (Cis-Di-(thiocyanato) bis (2,2'-bipyridyl)-4,4'-dicarboxylate) ruthenium (II)) owing to its good light absorber and charge transfer properties vis-à-vis to any other dyes [15]. A volatile electrolyte such as iodide/triiodide is commonly used which has a highly corrosive nature and good reaction with Platinum (Pt) based counter electrode [16, 17]. The photo-anode of DSSC is usually coated employing chemical route-based techniques like doctor blade and spin coating followed by high-temperature heat treatment [18–20].

DSSC can be useful for portable electronic devices, iPods, and solar lamps that work on the outdoor light source. The outdoor performance of the DSSC device was observed by many scientists in terms of the commercialization of DSSC. But the main factor that affects solar efficiency is a temperature that decreases the long-term stability of the device. To increases the stability of DSSC it was covered by plastic but appropriate results are not obtained. DSSC can easily work in a low-light condition or cloudy condition so these cells are a good option for building integrated photovoltaic cells (BIPV). However, DSSC also exhibits photoresponse/EQE concerning dye and electrolyte. Higher the EQE/IPCE means the photon absorbed by dye molecule is high therefore regeneration of electrolyte takes place and high efficiency of the respected device is observed. This chapter comprises a basic introduction to solar cells viz. principle of solar cells, and description of dye-sensitized solar cells as well the outdoor performance and stability along with photoresponse external quantum efficiency of the solar cell (EQE).

A solar cell directly converts solar energy into electrical energy by a physical process termed as "photovoltaic effect". The conversion of energy occurs without any intermediate process in certain semiconductor materials. In the photovoltaic effect, a semiconducting material generates charge carriers (electrons in conduction band and corresponding holes in valence band) when it is exposed by light where the light or solar energy and optical energy band gap of the exposed material are the important parameters. In the photoelectric effect, charge carriers are electrons while in the photovoltaic effect, charge carriers are both the electrons and holes. The photovoltaic effect was firstly discovered in 1839 by French Physicist Edmond Becquerel. During experimentation with wet cells, Becquerel noted that the voltage of the cell increased when its silver plates were exposed to the sunlight [21]. The solar cells are composed of different types of semiconductors where p-type and n-type layers are joined together to form a p-n junction (**Figure 1**). The junction between two types of semiconductors promotes an electrical field which is formed in the region of the junction as electrons move towards the positive p-side and holes towards the negative n-side. This generated field causes negatively charged carriers to move in one direction and positively charged carriers in opposite direction. On connecting it with the load, an electric current is produced in the circuit.

The sunlight is composed of photons which are the smallest energy bundles of electromagnetic radiation or energy. These photons can be absorbed by the absorber layer of the photovoltaic cell if the photons have energy (hυ) equal or greater than E_g and less than 2 E_g where E_g is the band gap of the layer concerned. When the light of a suitable wavelength is incident on these cells, energy from the photon is transferred to an atom of the semiconducting material in the p-n junction. Specifically, energy is transferred to the electrons in the material. This causes the electrons to jump to a higher energy level which is known as the conduction band. This leaves behind a "hole" in the valance band from which an electron is jumped up. This movement of the electron as a result of added energy creates two charge carriers viz. electrons in the conduction band and holes in the valence band. The asymmetric junction of different natures of semiconducting materials in the solar cell leads to the separation of these charge carriers (electron and holes) and establishes the built-in potential which impels these charge carriers towards the respective electrodes to contribute to electric current in the circuit.

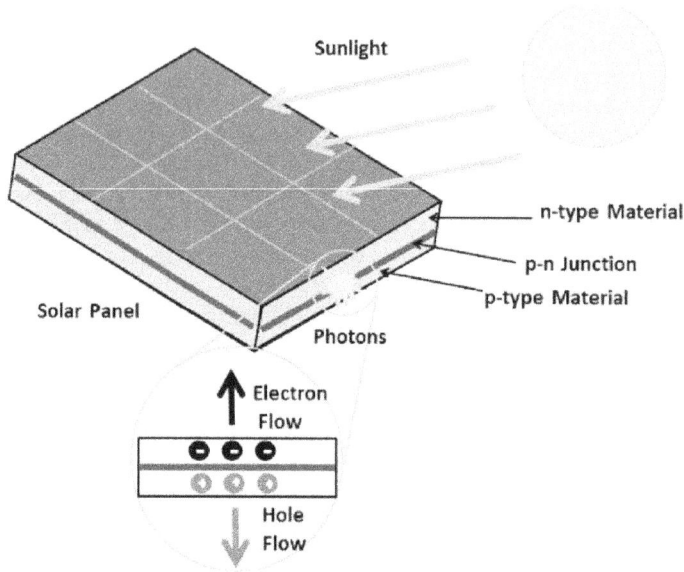

Figure 1.
A systematic presentation of the photovoltaic effect [22].

2. Dye-sensitized solar cells

As stated in the introduction part that the energy demand has increased the depletion of fossil fuels, therefore, the development of new skills which are based on renewable energy resources are spurred by world-leading scientists so that the upcoming new generation does not face any crisis related to the energy. Photovoltaic technology is eco-friendly and attractive among all renewable energy technologies. It directly converts sunlight into electrical energy, thus, it is broadly used for harvesting solar energy. The conventional Silicon-based solar cells are quite restricted because of their high cost, hence inexpensive, environmentally friendly, and simple fabrication process-based solar cells such as dye-sensitized solar cells (DSSCs) are used [23]. The dye-sensitized solar cells are comprised of a semiconducting material photo-anode, a counter electrode, an electrolyte, and a sensitizer (dye). DSSC can work in dark and cloudy conditions so it is an excellent candidate for indoor applications. O'Regan and Gratzel developed the first dye-sensitized solar in 1991 by colloidal nanoparticles of TiO_2 thin films which had an efficiency of 7.1%. The main aim of the present chapter is to introduce DSSC therefore, it is discussed in detail.

2.1 Device structure and working principle of dye-sensitized solar cell

A typical dye-sensitized solar cell is assembled in a sandwich-type structure. Generally, transparent conductive glass is used as a substrate for the deposition of nanocrystalline thin films of metal oxide. The metal oxide films are sensitized by absorbing dye molecules where dye is covalently attached to the surface of the photo-anode for generating the photoelectrons. An organic electrolyte solution that contains redox couple is used for collecting electrons at the surface of the counter electrode and regenerating dye molecules. A catalyst deposited on a conductive substrate is used as a counter electrode for the development of dye-sensitized solar cells [24]. The schematic representation of the device structure to a typical DSSC is shown in **Figure 2**.

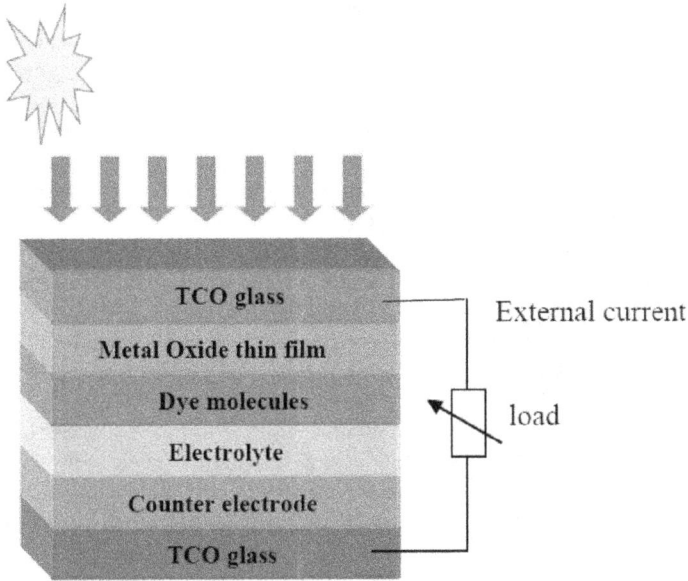

Figure 2.
Schematic device structure of a typical dye-sensitized solar cell (DSSC).

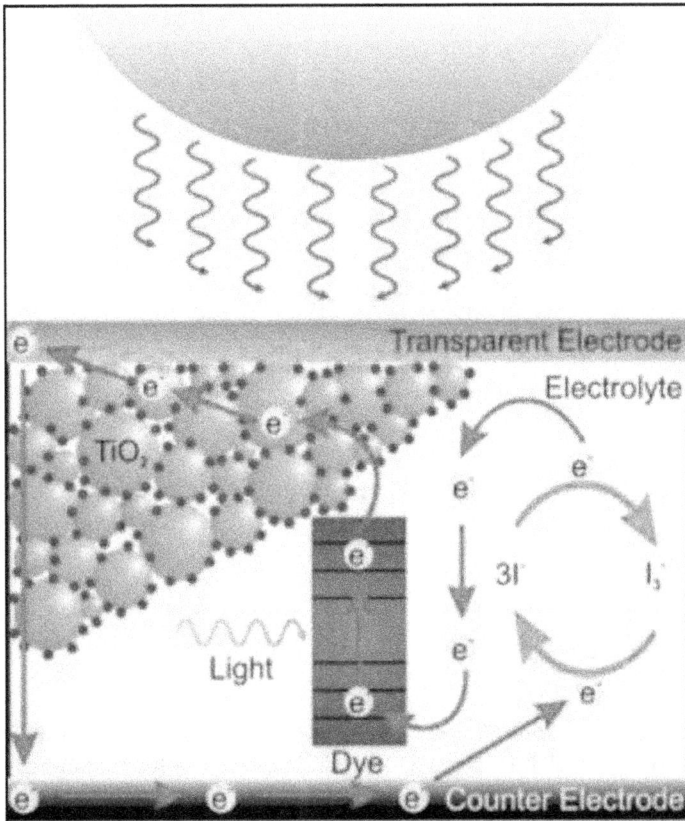

Figure 3.
A pictorial view of the operational principle of a typical dye-sensitized solar cell [26].

The absorption of irradiance and charge separation is quite different in the dye-sensitized solar cell as compared to the classical p-n junction solar cell [25]. An electron transfer process of sandwich-type dye-sensitized solar cells is systematically represented in **Figure 3**. The whole working process of the dye-sensitized solar cell is explained in three steps (1) Photo-excitation, (2) Transportation, and (3) Regeneration.

When the sunlight falls on a dye-sensitized solar cell device, then the present dye molecules on the surface of the TiO_2 layer (behaves like electron transport layer) absorb the incident photons and consequently excite the electrons. The excited electrons of dye which present above the conduction band of TiO_2 are immediately injected into the conduction band of TiO_2 and dye molecules get oxidized. At this stage, an electrochemical potential difference is generated between semiconductor oxide and electrolyte, and the electron density of TiO_2 also is increased due to charge carrier transfer from dye molecules to metal oxide.

Now, these electrons transfer from metal oxide to counter electrode through the external load where these electrons further transfer to the electrolyte. Herein, reduction of the electrolyte takes place by converting tri-iodide (I_3^-) into iodide (I^-). Regeneration of dye molecules is occurred by receiving electrons from iodide and simultaneous oxidation of iodide to tri-iodide happens. Regeneration of I^- is taken place by counter electrode so the whole cycle is regenerated. The flow of electrons through the external circuit generates electrical energy [27–29].

The chemical reactions that took place in the mechanism are given as below [30–33]:

2.1.1 The chemical reaction of dye-sensitized solar cell

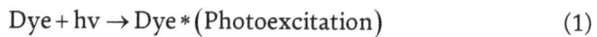

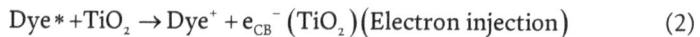

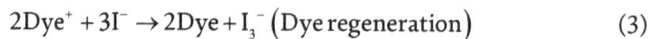

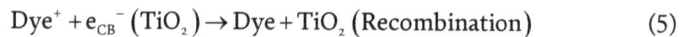

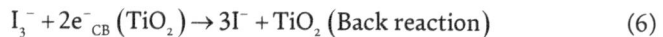

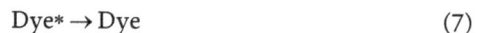

$$Dye + hv \rightarrow Dye * (Photoexcitation) \tag{1}$$

$$Dye * + TiO_2 \rightarrow Dye^+ + e_{CB}^- (TiO_2)(Electron\ injection) \tag{2}$$

$$2Dye^+ + 3I^- \rightarrow 2Dye + I_3^- (Dye\ regeneration) \tag{3}$$

$$I_3^- + 2e^-_{catalyst} \rightarrow 3I^- (Electrolyte\ regeneration) \tag{4}$$

$$Dye^+ + e_{CB}^- (TiO_2) \rightarrow Dye + TiO_2 (Recombination) \tag{5}$$

$$I_3^- + 2e^-_{CB}(TiO_2) \rightarrow 3I^- + TiO_2 (Back\ reaction) \tag{6}$$

$$Dye* \rightarrow Dye \tag{7}$$

2.2 Components of dye-sensitized solar cell

a. **Substrate:** Generally, a transparent conductive glass substrate is used for the fabrication of thin-film layers which could be employed as transparent conducting oxide substrates to develop a device. The transparent conducting oxide can be either Fluorine doped Tin oxide (FTO) or Indium doped Tin oxide (ITO) [34]. FTO substrate is usually applied for DSSC owing to good conduction property, stability, durability, and low toxicity. Besides the conductive glass substrates, the plastic materials, metal sheets, Ti foils are also useful substrates for DSSCs and according to the device, architecture may be designed.

b. **Photo-anode:** In DSSC, the photo-anode is a wide bandgap semiconducting material e.g. TiO_2, ZnO, SnO_2, ZrO_2, Nb_2O_5, Al_2O_3 are used as photo-anode for device development [35–41]. The main goal of these semiconductor materials is to absorb dye molecules and collection of the photo-excited electrons. Photo-anode materials should have a high surface area so absorption of dye molecules could be increased which eventually enhanced the power conversion efficiency of the solar cell device concerned. The crystallite size, porosity, microstructure, etc. play an important role to develop an efficient device for maximum harvesting of the incident irradiance. Typically, the DSSC photo-anode is prepared by conventional technique i.e. doctor blade but nowadays, many techniques are available which could be applied as per need and device architecture [42–46]. Doping of semiconductor material with suitable cation or anion also alters its optical energy bandgap, and post-deposition treatments like annealing affect the electrical, structural, and other relevant properties [47–49].

c. **Counter electrode:** The counter electrode (cathode) plays an important role in the regeneration of electrolytes by transporting electrons to the electrolyte which arrived externally from the circuit. Thus, the counter electrode should have good conductivity and catalytic activity. Platinum (Pt) is normally preferred to choose as a counter electrode for dye-sensitized solar cells [50]. The high cost and corrosion of Platinum limit its use and therefore, alternative options could be undertaken for counter electrodes. Carbon and conducting polymers (PEDOT) are also suitable materials due to their low cost, abundance, and adequate conductivity but their catalytic activity is lower as compared to the Platinum [51, 52]. Besides these, NiS/rGO, polypyrrole (PPy), $Co_{0.5}Ni_{0.5}Se/GN$, and WO_2 are utilized as counter electrodes for dye-sensitized solar cells [53–56].

d. **Electrolyte:** The function of electrolyte is to regenerate dye molecules and to work as conducting medium. Electrolyte plays important role in achieving higher efficiency of a solar cell. Based on the physical state, the electrolytes are classified into three main categories as a liquid electrolyte, quasi-solid electrolyte, and solid electrolyte [57]. As a liquid, triiodide/iodide (I_3^-/I^-) is mostly used as a redox couple because of the fast regeneration of the dye and slow recombination process in the dye-sensitized solar cell. Other electrolytes are also available like Br^-/Br_3^-, $SeCN^-/(SeCN)_2$, $SCN^-/(SCN)_2$, Co (II)/(III), Cu (I/II) etc. [58–60]. To overcome the problem of volatilization and leakage of liquid electrolytes, the quasi-solid and solid electrolytes are explored. Quasi solid electrolytes are organic liquid polymers that are converted into gel form by chemical and physical reactions that have cohesive nature and diffusive transport properties [61]. As solid electrolytes, mainly hole-transporting materials (HTM) are used viz. spiro-OMeTAD, CuSCN, CuI, P3HT, PEDOT, $CsSnI_3$ which can overcome the issue of leakage, corrosion, and salvation for DSSCs [62]. In HTM or semiconductors, charge transportation takes place via electrons or holes while in electrolytes, it takes place through ions.

e. **Sensitizer:** Dyes play the important role of photo-sensitizer in DSSCs where a self-assembled layer of dye is anchored on the surface of the photo-anode. When sunlight strikes on dye molecules then these dye molecules absorb photons and consequently, the photoexcitation of electrons occurs which injects electrons into the conduction band of the photo-anode. Based on the composition used in dye, it is classified into three main categories viz. metal complex, metal-free organic complex, and natural sensitizer [63]. Ruthenium-based sensitizers are remarkable for achieving higher efficiency in dye-sensitized solar cells [64, 65].

3. Outdoor performance and stability of dye-sensitized solar cell

In outdoor conditions the main factor that affects efficiency is temperature. In this section different solar radiation illumination was discussed. The stability of dye-sensitized solar cell is mainly influenced by electrolyte, liquid electrolyte exhibit higher efficiency but the volatile nature of liquid electrolyte degrades solar cell and therefore stability of the cell reduces.

Yuan et al. [66] tested outdoor application of DSSC for Building Integrated Photovoltaics (BIPV) application where a time duration of four years was taken into consideration. Here dye Z991 and Z907 were used for cell fabrication where the first one harnesses 15% more electricity over the later one for two years. Given the stability of the device, the efficiency of dye Z991 based solar cell decreases to 17% for the initial two years thereafter efficiency remains stable for the remaining two years. Moreover, the Z907 based DSSC device is out of the use or degrade after four years in outdoor application. The stability of Z991 over Z907 is due to the presence of thiophene moieties in Z991 i.e. responsible for better energy harvesting and thermal stability. When the solar irradiance increase there is no linearly incremented in electricity generation for irradiance of lower than 20 Wh.

Kato et al. [67] synthesized dye-sensitized solar cells with N719 dye, TiO_2, and carbon counter electrode and tested durability test in the outdoor working condition for a time duration of 2.5 years. The DSSC modules were developed monolithically series interconnected on the TCO substrate and covered by a waterproof cover. Before the exposure to sunlight, the device reveals 0.32 and 0.71 suns from the current–voltage curve and power-voltage curve. During the stability test voltage was approximately kept around 1.6 V. The solar parameters such as JSC fall for 5 months thereafter it remains constant for left years and efficiency decreases/degrades subsequently decrement in VOC and FF. Additionally, EIS reveals exposure of cell in outdoor increases the Nernst impedance of triiodide and Raman spectra also reveals increment in luminescent ingredients of electrolyte, therefore, V_{OC} and FF decreases in outdoor condition. Berginc et al. [68] outdoor exposed ionic liquid-based dye-sensitized solar cell for 7 months in solar radiation of $906kWh/m^2$. In the summers maximum V_{OC} is obtained in the early morning and on an autumn day when days are shorter and temperature is lower that time J_{SC} of cell increases. Park et al. [69] observed the change in film thickness effects J-V curve (**Figure 4**) of TiO_2 based solar cell under 1 sunlight intensity. On increasing the thickness, the J^{SC} of the cell increases from 6.6 to 10.7 mA/cm^2 i.e. about 62% whereas the fall down in V_{OC} is 759 to 727 mV due to increases in surface area that accounts for more dye molecule absorbing.

Asghar et al. [70] developed dye-sensitized solar cell and tested in outdoor condition as well comparison with silicon cell was carried out. Here the lower irradiance and higher temperature are suitable for DSSC, at these parameter DSSC harvest more energy instead of silicon solar cell. The efficiency of DSSC decreases as time duration increases. Moreover, the device that was fabricated by employing MPN as an electrolyte degrades fast whereas ionic liquid-based devices are more stable and constant efficiency was observed for two months then degradation initiates. The thermal influence of dye-sensitized solar cells was studied by Matsui et al. [71] where the current collecting study was done. When the temperature was maintained at around 85° C leakage of ionic liquid does not occur but the long-term stability of the device is strongly affected by moisture. Therefore double-sealed

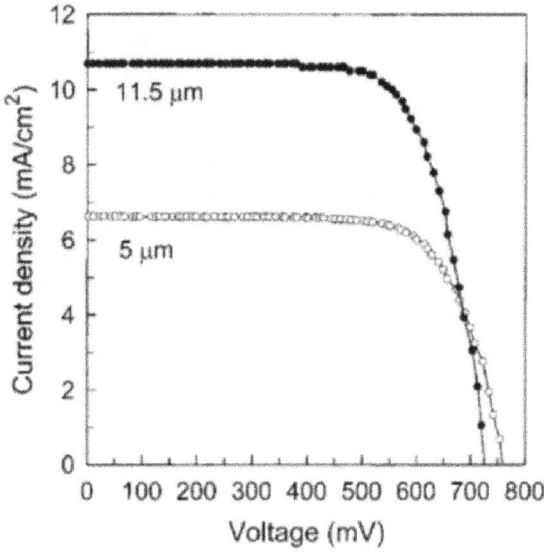

Figure 4.
*The effect of film thickness on the J-V curve of DSSC. Reprinted with permission from ref. [69] copyright
(2000) American Chemical Society.*

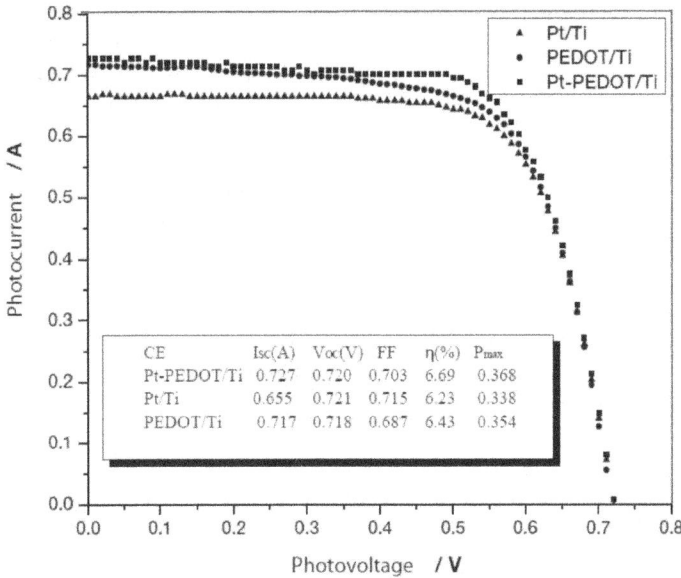

CE	Isc(A)	Voc(V)	FF	η(%)	Pmax
Pt-PEDOT/Ti	0.727	0.720	0.703	6.69	0.368
Pt/Ti	0.655	0.721	0.715	6.23	0.338
PEDOT/Ti	0.717	0.718	0.687	6.43	0.354

Figure 5.
*The J-V curve for DSSC with the different counter electrodes. Reprinted with permission from ref. [73]
copyright (2012) John Wiley and Sons.*

package for the device was invented and a test on substrate size of 50 mm × 50 mm
was used where 85° C temperature was maintained for 1000 hours and stability
was observed. Bella et al. [72] designed fluoropolymer and rare elements-free
light shifting coating systems for dye-sensitized solar cell devices. The introduction of fluorescent species in DSSC downshifts UV photons into visible light that
significantly improves PV efficiency by 60%. The improvement in efficiency is

accountable for improvement in photon flux i.e. caused by the introduction of luminescent agent that results from nanometric light shifting in organic dyes. Now the outdoor long-term stability was measured for 3 months where the introduction of a light shifting agent preserves the power conversion efficiency of the solar cell.

Freitag et al. [74] demonstrated dye-sensitized solar cells with dye D35 and XY1 were copper-based redox electrolyte is used. At the AM of 1.5 G, the observed PCE is 11.3% and under 1000 lux indoor condition it achieves 28.9%. The obtained results point out DSSC are suitable for ambient light condition. Mehmood et al. [75] constructed DSSC with an organic photosensitizer. The PCE of the cell was 2.58% at 25°C in air mass of 1.5 G and illumination of 100 mW/cm^2. The increment in temperature falls down the efficiency of this solar cell it is stable up to 35°C. Wu et al. [73] demonstrated dye-sensitized solar cells with an area of 100 cm^2 and lightweight based on Ti substrates. Here PEDOT counter electrode is used which is having good transparency and electrocatalytic activity. The J-V curve (**Figure 5**) reveals the current density (I_{SC}) of PEDOT-Pt/Ti is higher. The photoconversion efficiency was achieved about 6.69% and in an outdoor condition of solar radiation of 55 mW cm^{-2} 0.368 W power output was observed.

4. Spectral response/external quantum efficiency (EQE) response/ incident photon-to-current conversion efficiency (IPCE)

Generally, dye-sensitized solar cell photoresponse for a given incident wavelength of light and the result is depicted in form of varying wavelength and percentage of IPCE. When the current is generated through the response of photon that time characteristics peak appears at a particular wavelength. It is the ratio of generated electrons to the incident photons. Moreover, IPCE depends upon the yield of electron transfer and light-harvesting efficiency that causes quantum charge injection and electron quantum efficiency in the present external circuit of the device. In the case of DSSC, the measurement of IPCE clears that dye is well linked to photoanode and electrolyte. When incident photons are exposed on DSSC that time dye uptake electrons from photoanode and create electron–hole pair and holes are transmitted to the electrolyte.

The generation of photocurrent i.e. dependent on wavelength is known as external quantum efficiency (EQE) where AC and DC mode is used for the generation of the beam. In the case of DC mode irradiation of monochromatic beam on a sample is continuously carried out for 3 sec so electrons reach to steady-state. In AC mode monochromatic light is chopped by shutter and illumination of bias light on a sample is carried out. Jeong et al. [76] measured EQE of DSSC and tandem cell (DSSC/CIGS) in DC mode. The EQE spectra reveal in the wavelength range of 300–800 nm EQE of DSSC was observed and for tandem cells, EQE spectra are almost similar to DSSC. When Berginc et al. [68] DSSC was exposed to outdoor conditions for seven months, the EQE of the solar cell was measured. The peak at 360 nm is accountable for absorption in the TiO$_2$ layer, 380 nm for change in I3- and at 450 nm for dye molecules degradation.

Kubo et al. [77] developed a tandem structure-based solar cell that improves the photocurrent of dye-sensitized solar cells. The IPCE of tandem solar cells is relatively outstanding to single cells. Tandem solar cell has elevated solar response (good external quantum efficiency), photocurrent and conversion efficiency from single-cell as well lower VOC and higher FF was also observed. Park et al. [69] prepared homogeneous, crack-free, and rod-shaped rutile TiO2 thin films with having a thickness of 12 μm. The measurement of IPCE (incident photon-to-current efficiency) till 600 nm wavelength indicated that a significant amount of light was absorbed very fast in few microns but at higher wavelength, the increment in IPCE was directly proportional to the film thickness see **Figure 6**. Rutile and anatase films were compared having similar thickness where photocurrent of rutile based solar cell was 30% lower vis-a-vis to the anatase phase owing to the less amount of absorbed dye, small surface area, and transportation of electrons was also slow for rutile thin film-based solar cells. Lepikko et al. [78] tested outdoor performance of DSSC for 1000 h in 1 sun. The efficiency and fill factor of cell rise in outdoor condition i.e. just double of indoor condition well-remaining of solar irradiance. The IPCE decreases about 30% during testing of the cell this is due to photodegradation of electrolyte see **Figure 7**.

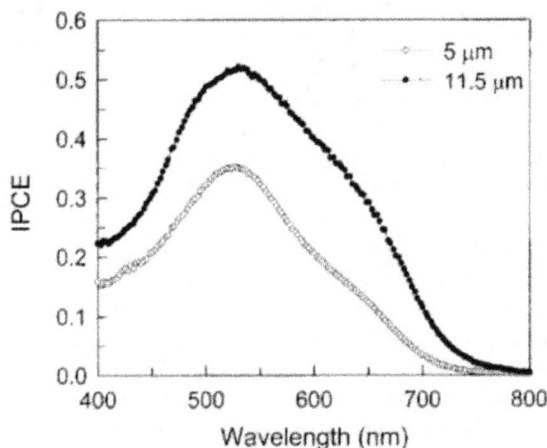

Figure 6.
The effect of a film thickness of TiO₂ on IPCE value of a solar cell. Reprinted with permission from ref. [69] copyright (2000) American Chemical Society.

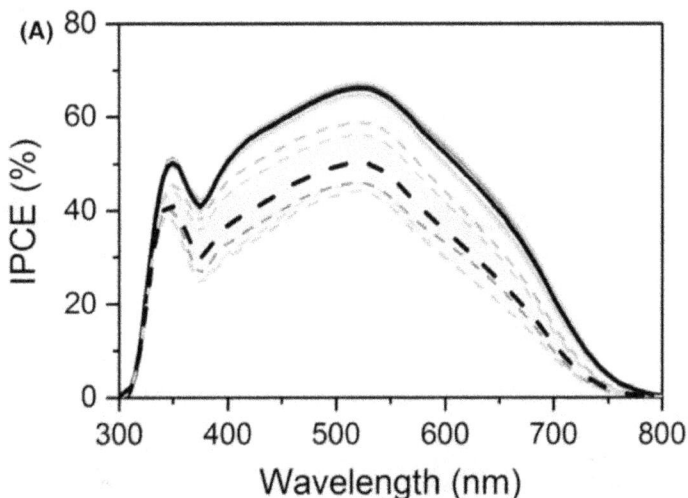

Figure 7.
The IPCE curve of DSSC in harsh northern outdoor conditions. Reprinted with permission from ref. [78] copyright (2018) John Wiley and Sons.

Roy et al. [79] studied the annealing of TiO2 nanotubes at 450°C for 30 minutes where the amorphous phase was converted into anatase. Post annealing and the TiCl4 treatments were carried in a closed vessel at 70°C for 30 min. SEM image of TiO2 nanotubes treated with TiCl4 confirmed uniform decoration with TiO2 nanoparticles and IPCE of the decorated samples was found 66% with a conversion efficiency of 3.8%. The ultrathin nanosheets of SnO2 were introduced as photo-anode in dye-sensitized solar cells for improvement in photoconversion efficiency by Xing et al. [80]. The nanosheets were developed by hydrothermal method and screen printed over FTO substrates, then a coating of TiO2 on SnO2 was performed to solve the problem of lower open-circuit voltage. The diffraction peak in XRD patterns revealed to the tetragonal rutile like SnO2 and FESEM images displayed a 3D flowerlike structure. HRTEM images of nanosheets showed lattice fringes over the entire surface. The efficiency of the devices using SnO2 NSs-TiO2 was 1.79% and IPCE was 35% which was much higher vis-a-vis the devices made up of SnO2 nanoparticles i.e. revealed by **Figure 8**.

Kumara et al. [81] employed natural dyes obtained from Ixora sp. (IX) and Canarium odontophyllum (CMB) which mainly contained anthocyanin that was used to improve the performance of DSSCs. The layered co-sensitization of dyes was carried out by firstly immersing TiO2 electrode in CMB extract followed by de-adsorption and then again immersed in second sensitizer IX for adsorption. The absorption spectrum of the co-sensitized electrode was increased as compared to the individual and mixture sensitized and similar results were obtained in IPCE measurement. The photovoltaic properties of the co-sensitized electrode were obtained under irradiance of 1000 W/m2 with a short circuit current density of 9.80 mA/cm2, VOC of 343 mV, fill factor of 0.46, and photoconversion efficiency of 1.55%. Gupta et al. [82] developed Cu/S co-doped TiO_2 as a photoanode for dye-sensitized solar cells. Here undoped TiO_2 exhibits about 70.02% of IPCE whereas it increases further on codoping with Cu/S. 0.1% Cu/S exhibits 73.65% of IPCE and on increment, the 0.3% Cu/S exhibits 82.98% of IPCE at a wavelength of 530 nm.

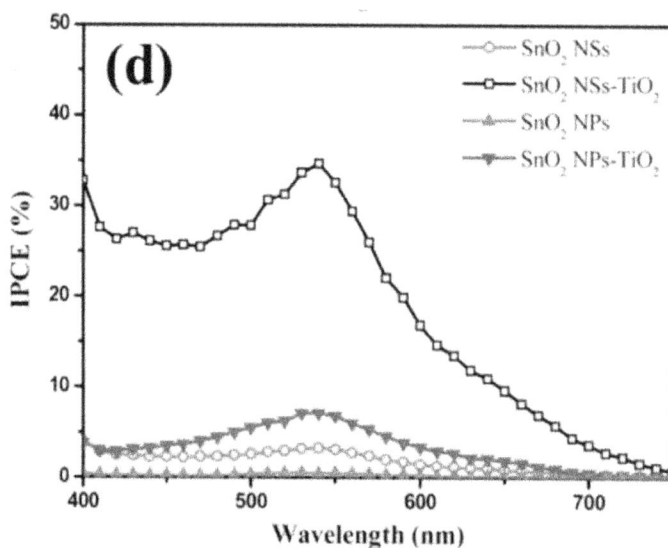

Figure 8.
IPCE curve of DSSC with different photoelectrode. Reprinted with permission from ref. [80] copyright (2012) American Chemical Society.

The improvement in IPCE is accountable due to the small size of particles and enhancement in short circuit current density (J_{SC}).

Patni et al. [83] fabricated dye-sensitized solar cells with natural dyes. The natural dyes were used are anthocyanin, betalain, and chlorophyll obtained from the extracts of Roselle spinach beetroot respectively. At the wavelength of 430 nm 6.21% IPCE was observed for anthocyanin dye and at 530 nm 9.9% of IPCE was measured for betalain and 6.1% IPCE was observed for chlorophyll-based dye at a wavelength of 660 nm. The blending or mixing of dye improves the IPCE. Wood et al. [84] reported the IPCE for different dye i.e. The cationic 1-hexyl-2,3,3-3H indolium acceptor dye (CAD3) dye exhibit IPCE of 50% and bodipy dye, it is 53% and for P1 54% was observed where the p-type dye-sensitized solar cell was fabricated. This chapter comprised literature on the solar response of DSSC on exposure of induced photons. Different dyes exhibit a variation in photon-to-current conversion efficiency. The higher the IPCE means the efficiency of the cell is a good and better amount of energy can be harvested by solar cell.

5. Conclusion

The dye-sensitized solar cell technology has an impact on the PV market owing to easy fabrication, cost, chemical stability, availability of chemicals, and good power conversion efficiency. In this chapter, we discussed the introduction of solar cells with working principles, complete elaboration of dye-sensitized solar cells, and outdoor performance and stability in different solar irradiations. Outdoor performance is affected by the temperature because on exposure to sunlight the temperature raise degrades the electrolyte and therefore stability and performance of the device decreases. Moreover, on rainy days the chances of degradation are increasing due to water or moisture, therefore, coating/layer of a suitable material is carried out over the solar cell this also increases the long-term stability of the device. The IPCE of solar cells initially is higher but with time duration it falls due to cell degradation or leaking and for a long time it stabilizes without so many changes. This chapter emphasizes the efficiency of DSSC when it exposes the outdoor and solar response of DSSC.

Acknowledgements

The authors thank Mohanlal Sukhadia University, Udaipur. Reema Agarwal greatly acknowledges CSIR, New Delhi for financial support through Senior Research Fellowship (File No.: 09/172(0090)/2019-EMR-I).

Conflict of interest

The authors declare no conflict of interest.

Author details

Reema Agarwal, Yogeshwari Vyas, Priyanka Chundawat, Dharmendra
and Chetna Ameta[*]
Department of Chemistry, Mohanlal Sukhadia University, Udaipur, India

[*]Address all correspondence to: chetna.ameta@yahoo.com

IntechOpen

References

[1] Painter DS. Oil and geopolitics: The oil crises of the 1970s and the cold war. Historical Social Research. 2014;39:186-208. DOI: 10.12759/hsr.39.2014.4. 186-208.

[2] https://www.popularmechanics.com/science/energy/a21756137/renewables-50-percent-energy-2050/ [Internet].

[3] Green MA, Dunlop ED, Dean HL, Jochen HE, Masahiro Y, Anita WYH. Mint: Solar cell efficiency tables (version 54). Progress in Photovoltaics. 2019;27:565-575. DOI:10.1002/pip.3171

[4] Dréon J, Jeangros Q, Cattin J, Haschke J, Antognini L, Ballif C, Boccard M. Mint: 23.5%-efficient silicon heterojunction silicon solar cell using molybdenum oxide as hole-selective contact. Nano Energy. 2020;70:104495. DOI: 10.1016/j.nanoen.2020.104495

[5] Carlson DE, Wronski CR. Mint: Amorphous silicon solar cell. Applied Physics Letters. 1976;28:671-673. DOI: 10.1063/1.88617

[6] Chu TL, Chu SS. Mint: Recent progress in thin-film cadmium telluride solar cells. Progress in Photovoltaics. 1993;1:31-42. DOI: 10.1002/pip.4670010105

[7] Kazmerski LL, White FR, Morgan GK. Mint: Thin-film CuInSe$_2$/CdS heterojunction solar cells. Applied Physics Letters. 1976;29:268-269. DOI: 10.1063/1.89041

[8] Ranabhat K, Patrikeev L, Revina AA, Andrianov K, Lapshinsky V, Sofronova E. Mint: An introduction to solar cell technology. Journal of Applied Engineering Science. 2016;14:481-491. DOI 10.5937/jaes14-10879

[9] Halme J. Dye-sensitized nano-structured and organic photovoltaic cells: technical review and preliminary tests [thesis]. Finland: Helsinki University of Technology; 2002.

[10] Tang YB, Lee CS, Xu J, Liu ZT, Chen ZH, He Z, Cao YL, Yuan G, Song H, Chen L, Luo L, Cheng HM, Zhang WJ, Bello I, Lee ST. Mint: Incorporation of graphenes in nanostructured TiO$_2$ films via molecular grafting for dye-sensitized solar cell application. ACS Nano. 2010;4:3482-3488. DOI: 10.1021/nn100449w

[11] Gratzel M. Mint: Dye-sensitized solar cells. Journal of Photochemistry and Photobiology C: Photochemistry Reviews. 2003;4:145-153. DOI: 10.1016/S1389-5567(03)00026-1

[12] Kakiage K, Aoyama Y, Yano T, Oya K, Fujisawa J-I, Hanaya M. Mint: Highly-efficient dye-sensitized solar cells with collaborative sensitization by silyl-anchor and carboxy anchor dyes. Chemical Communications. 2015;51:15894-15897. DOI: 10.1039/C5CC06759F

[13] Graetzel M, Janssen RAJ, Mitzi DB, Sargent EH. Mint: Materials interface engineering for solution-processed photovoltaics. Nature. 2012;488: 304-312. DOI: 10.1038/nature11476

[14] Li T-Y, Su C, Akula SB, Sun W-G, Chien H-M, Li W-R. Mint: New pyridinium ylide dyes for dye sensitized solar cell applications. Organic Letters. 2016;18:3386-3389. DOI: 10.1021/acs.orglett.6b01539

[15] Wei L, Chen S, Yang Y, Dong Y, Song W, Fan R. Mint: Reduced graphene oxide modified TiO$_2$ semiconductor materials for dye-sensitized solar cells. RSC Advances. 2016;6:100866-100875. DOI: 10.1039/C6RA22112B

[16] Liu Y, Yun S, Zhou X, Hou Y, Zhang T, Li J, Hagfeldt A. Mint: Intrinsic origin of superior catalytic properties of tungsten-based catalysts in dye-sensitized solar cells. Electrochimica Acta. 2017;242:390-399. DOI: 10.1016/j.electacta.2017.04.176

[17] Hashmi SG, Özkan M, Halme J, Zakeeruddin SM, Paltakari J, Grätzel M, Lund PD. Mint: Dye-sensitized solar cells with inkjet-printed dyes. Energy & Environmental Science. 2016;9:2453-2462. DOI: 10.1039/C6EE00826G

[18] Bhand S, Chadar D, Pawar K, Naushad M, Pathan H, Salunke-Gawali S. Mint: Benzo [α] phenothiazine sensitized ZrO_2 based dye sensitized solar cell. Journal of Materials Science: Materials in Electronics. 2018;29:1034-1041. DOI: 10.1007/s10854-017-8003-2

[19] Ünlü B, Çakar S, Özacar M. Mint: The effects of metal doped TiO_2 and dithizone-metal complexes on DSSCs performance. Solar Energy. 2018;166:441-449. DOI: 10.1016/j.solener.2018.03.064

[20] Ruhane TA, Islama MT, Rahaman MS, Bhuiyan MMH, Islam JMM, Bhuiyan TI, Khan KA, Khan MA. Mint: Impact of photo electrode thickness and annealing temperature on natural dye sensitized solar cell. Sustainable Energy Techno-logies and Assessments. 2017;20:72-77. DOI: 10.1016/j.seta.2017.01.012

[21] Boyle G. Renewable Energy: Power for a Sustainable Future. UK:Oxford University Press;1996. 479 p. DOI:

[22] https://energyeducation.ca/wiki/images/1/11/Photovoltaiceffect.png [Internet].

[23] Gong J, Sumathy K, Qiao Q, Zhou Z. Mint: Review on dye-sensitized solar cells (DSSCs): Advanced techniques and research trends. Renewable and Sustainable Energy Reviews. 2017;68:234-246. DOI: 10.1016/j.rser.2016.09.097

[24] Lee C-P, Ho K-C. Mint: Poly (ionic liquid)s for dye-sensitized solar cells: A mini-review. European Polymer Journal. 2018;108:420-428. DOI: 10.1016/j.eurpolymj.2018.09.022

[25] Sengupta D, Das P, Mondal B, Mukherjee K. Mint: Effects of doping, morphology and film-thickness of photo-anode materials for dye sensitized solar cell application – A review. Renewable and Sustainable Energy Reviews. 2016;60:356-376. DOI: 10.1016/j.rser.2016.01.104

[26] https://en.wikipedia.org/wiki/Dye-sensitized_solar_cell [Internet].

[27] Calogero G, Bartolotta A, Marco GD, Carlo AD, Bonaccorso F. Mint: Vegetable-based dye-sensitized solar cells. Chemical Society Reviews. 2015;44:3244-3294. DOI: 10.1039/C4CS00309H

[28] Roslan N, Ya'acob ME, Radzi MAM, Hashimoto Y, Jamaludin D, Chen G. Mint: Dye sensitized solar cell (DSSC) greenhouse shading: New insights for solar radiation manipulation. Renewable and Sustainable Energy Reviews. 2018; 92:171-186. DOI: 10.1016/j.rser.2018. 04.095

[29] Richhariya G, Kumar A, Tekasakul P, Gupta B. Mint: Natural dyes for dye sensitized solar cell: A review. Renewable and Sustainable Energy Reviews. 2017;69:705-718. DOI: 10.1016/j.rser.2016.11.198

[30] Deb Nath NC, Lee J-J. Mint: Binary redox electrolytes used in dye-sensitized solar cells. Journal of Industrial and Engineering Chemistry. 2019;78:53-65. DOI: 10.1016/j.jiec.2019.05.018

[31] Kumara NTRN, Limb A, Lim CM, Petra MI, Ekanayake P. Mint: Recent progress and utilization of natural pigments in dye sensitized solar cells: A review. Renewable and Sustainable Energy Reviews. 2017;78:301-317. DOI: 10.1016/j.rser.2017.04.075

[32] Mehmood U, Al-Ahmed A, Al-Sulaiman FA, Malik MI, Shehzad F, Khan AUH. Mint: Effect of temperature on the photovoltaic performance and stability of solid-state dye-sensitized solar

cells: A review. Renewable and Sustainable Energy Reviews. 2017;79: 946-959. DOI: 10.1016/j.rser.2017.05.114

[33] Iqbal MZ, Khan S. Mint: Progress in the performance of dye sensitized solar cells by incorporating cost effective counter electrodes. Solar Energy. 2018;160:130-152. DOI: 10.1016/j. solener.2017.11.060

[34] Sima S, Grigoriu C, Antohe S. Mint: Comparison of the dye-sensitized solar cells performances based on transparent conductive ITO and FTO. Thin Solid Films. 2010;519:595-597. DOI: 10.1016/j. tsf.2010.07.002

[35] Peng T, Xu J, Chen R. Mint: A novel multilayer brookite TiO$_2$ electrode for improved performance of pure brookite-based dye sensitized solar cells. Chemical Physics Letters. 2020;738:136902. DOI: 10.1016/j.cplett.2019.136902

[36] Siwatch S, Kundu V, Kumar A, Kumar S. Mint: Role of surfactant in optimization of 3D ZnO floret as photo-anode for dye sensitized solar cell. Applied Nanoscience. 2020;10:1035-1044. DOI: 10.1007/s13204-019-01216-w

[37] Kavan L, Zivcova ZV, Zlamalova M, Zakeeruddin SM, Grätzel M. Mint: Electron-selective layers for dye-sensitized solar cells based on TiO$_2$ and SnO$_2$. The Journal of Physical Chemistry C. 2020;124:6512-6521. DOI: 10.1021/acs.jpcc.9b11883

[38] Bhalekar VP, Baviskar PK, Prasad B, Beedri NI, Kadam VS, Pathan HM. Mint: Lead sulphide sensitized ZrO$_2$ photo-anode for solar cell application with MoO$_3$ as a counter electrode. Chemical Physics Letters. 2017;689:15-18. DOI: 10.1016/j.cplett.2017.10.001

[39] Ghosh R, Brennaman MK, Uher T, Ok M-R, Samulski ET, McNeil LE, Meyer TJ, Lopez R. Mint: Nanoforest Nb$_2$O$_5$ photo-anodes for dye-sensitized solar cells by pulsed laser deposition. ACS Applied Materials & Interfaces. 2011;3:3929-3935. DOI: 10.1021/ am200805x

[40] Liu M, Yang J, Feng S, Zhu H, Zhang J, Li G, Peng J. Mint: Composite photo-anodes of Zn$_2$SnO$_4$ nanoparticles modified SnO$_2$ hierarchical microspheres for dye-sensitized solar cells. Materials Letters. 2012;76:215-218. DOI: 10.1016/j. matlet.2012.02.110

[41] Prasittichai C, Hupp JT. Mint: Surface modification of SnO$_2$ photoelectrodes in dye-sensitized solar cells: Significant improvements in photovoltage via Al$_2$O$_3$ atomic layer deposition. The Journal of Physical Chemistry Letters. 2010;1:1611-1615. DOI: 10.1021/jz100361f

[42] Okuya M, Nakade K, Kaneko S. Mint: Porous TiO$_2$ thin films synthesized by a spray pyrolysis deposition (SPD) technique and their application to dye-sensitized solar cells. Solar Energy Materials and Solar Cells. 2002;70: 425-435. DOI: 10.1016/S0927-0248(01) 00033-2

[43] Kim GS, Seo H-K, Godble VP, Kim YS, Yang OB, Shin HS. Mint: Electrophoretic deposition of titanate nanotubes from commercial titania nanoparticles: Application to dye-sensitized solar cells. Electrochemistry Communications. 2006;8:961-966. DOI: 10.1016/j.elecom.2006.03.037

[44] Li Y, Ma L, Yoo Y, Wang G, Zhang X, Ko MJ. Mint: Atomic layer deposition: A versatile method to enhance TiO$_2$ nanoparticles interconnection of dye-sensitized solar cell at low temperature. Journal of Industrial and Engineering Chemistry. 2019;73:351-356. DOI: 10.1016/j.jiec.2019.02.006

[45] Merazga A, Al-Subai F, Albaradi AM, Badawi A, Jaber AY, Alghamdic AAB. Mint: Effect of sol–gel MgO spin-coating on the performance of TiO$_2$-based dye-sensitized solar cells.

Materials Science in Semiconductor Processing. 2016;41:114-120. DOI: 10.1016/j.mssp.2015.08.026

[46] Li K, Wang Y, Sun Y, Yuan C. Mint: Preparation of nanocrystalline TiO_2 electrode by layer-by-layer screen printing and its application in dye-sensitized solar cell. Materials Science and Engineering: B. 2010;175:44-47. DOI: 10.1016/j.mseb.2010.06.019

[47] Chatterjee S, Webre WA, Patra S, Rout B, Glass GA, D'Souza F, Chatterjee S. Mint: Achievement of superior efficiency of TiO_2 nanorod-nanoparticle composite photo-anode in dye sensitized solar cell. Journal of Alloys and Compdounds. 2020;826:154188. DOI: 10.1016/j.jallcom.2020.154188

[48] Pandey P, Parra MR, Haque FZ, Kurchania R. Mint: Effects of annealing temperature optimization on the efficiency of ZnO nanoparticles photo-anode based dye sensitized solar cells. Journal of Materials Science: Materials in Electronics. 2017;28:1537-1545. DOI: 10.1007/s10854-016-5693-9

[49] Sadikin SN, Rahman MYA, Umar AA. Mint: Influence of annealing temperature of ZnS-coated TiO_2 films on the performance of dye-sensitized solar cells. Optik. 2020;211:16464. DOI: 10.1016/j.ijleo.2020.164644

[50] Bao Z, Xie H, Rao j, Chen L, Wei Y, Li H, Zhou X. Mint: High performance of Pt/TiO_2-nanotubes/Ti mesh electrode and its application in flexible dye-sensitized solar cell. Materials Letters. 2014;124:158-160. DOI: 10.1016/j.matlet.2014.03.041

[51] Wang W, Liu Y, Zhong YJ, Wang L, Zhou W, Wang S, Tadé MO, Shao Z. Mint: Rational design of $LaNiO_3$/Carbon composites as outstanding platinum-free photocathodes in dye-sensitized solar cells with enhanced catalysis for the triiodide reduction reaction.

Solar RRL. 2017;1:1700074. DOI: 10.1002/solr.201700074

[52] Sudhagar P, Nagarajan S, Lee Y-G, Song D, Son T, Cho W, Heo M, Lee K, Won J, Kang YS. Mint: Synergistic catalytic effect of a composite (CoS/PEDOT:PSS) counter electrode on triiodide reduction in dye-sensitized solar cells. ACS Applied Materials & Interfaces. 2011;3:1838-1843. DOI: 10.1021/am2003735

[53] Sarkar A, Chakraborty AK, Bera S. Mint: NiS/rGO nanohybrid: An excellent counter electrode for dye sensitized solar cell. Solar Energy Materials and Solar Cells. 2018;182:314-320. DOI: 10.1016/j.solmat.2018.03.026

[54] Bu C, Tai Q, Liu Y, Guo S, Zhao X. Mint: A transparent and stable polypyrrole counter electrode for dye-sensitized solar cell. Journal of Power Sources. 2013;221:78-83. DOI: 10.1016/j.jpowsour.2012.07.117

[55] Murugadoss V, Panneerselvam P, Yan C, Guo Z, Angaiah S. Mint: A simple one-step hydrothermal synthesis of cobalt-nickel selenide/graphene nanohybrid as an advanced platinum free counter electrode for dye sensitized solar cell. Electrochimica Acta. 2019; 312:157-167. DOI: 10.1016/j.electacta.2019.04.142

[56] Wu M, Lin X, Hagfeldt A Ma T. Mint: A novel catalyst of WO_2 nanorod for the counter electrode of dye-sensitized solar cells. Chemical Communications. 2011;47:4535-4537. DOI: 10.1039/C1CC10638D

[57] Wu J, Lan Z, Lin J, Huang M, Huang Y, Fan L, Luo G. Mint: Electrolytes in dye-sensitized solar cells. Chemical Reviews. 2015;115:2136-2173. DOI: 10.1021/cr400675m

[58] Kakiage K, Tokutome T, Iwamoto S, Kyomen T Hanaya M. Mint: Fabrication of a dye-sensitized solar cell containing

a Mg-doped TiO_2 electrode and a Br_3^-/ Br^- redox mediator with a high open-circuit photovoltage of 1.21 V. Chemical Communications. 2013;49:179-180. DOI: 10.1039/C2CC36873K

[59] Powar S, Daeneke T, Ma MT, Fu D, Duffy NW, Götz G, Weidelener M, Mishra A, Bäuerle P, Spiccia L, Bach U. Mint: Highly efficient p-type dye-sensitized solar cells based on Tris (1,2-diaminoethane) Cobalt (II)/(III) electrolytes. Angewandte Chemie. 2012;52:602-605. DOI: 10.1002/anie.201206219

[60] Freitag M, Giordano F, Yang W, Pazoki M, Hao Y, Zietz B, Grätzel M, Hagfeldt A, Boschloo G. Mint: Copper phenanthroline as a fast and high-performance redox mediator for dye-sensitized solar cells. The Journal of Physical Chemistry C. 2016;120:9595-9603. DOI: 10.1021/acs.jpcc.6b01658

[61] Vinoth S, Kanimozhi G, Narsimulu D, Kumar H, Srinadhu ES, Satyanarayana N. Mint: Ionic relaxation of electrospun nanocomposite polymer-blend quasi-solid electrolyte for high photovoltaic performance of dye-sensitized solar cells. Materials Chemistry and Physics. 2020;250:122945. DOI: 10.1016/j.matchemphys.2020.122945

[62] Hsu C-Y, Chen Y-C, Lin RY-Y, Ho K-C, Lin JT. Mint: Solid-state dye-sensitized solar cells based on spirofluorene (spiro-OMeTAD) and arylamines as hole transporting materials. Physical Chemistry Chemical Physics. 2012;14:14099-14109. DOI: 10.1039/C2CP41326D

[63] Mehrabian M, Dalir S. Mint: Numerical simulation of highly efficient dye sensitized solar cell by replacing the liquid electrolyte with a semiconductor solid layer. Optik. 2018;169:214-223. DOI: 10.1016/j.ijleo.2018.05.059

[64] Carlo GD, Biroli AO, Tessore F, Caramori S, Pizzotti M. Mint:

β-Substituted Zn^{II} porphyrins as dyes for DSSC: A possible approach to photovoltaic windows. Coordination Chemistry Reviews. 2018;358:153-177. DOI: 10.1016/j.ccr.2017.12.012

[65] Hug H, Bader M, Mair P, Glatzel T. Mint: Biophotovoltaics: Natural pigments in dye-sensitized solar cells. Applied Energy. 2014;115:216-225. DOI: 10.1016/j.apenergy.2013.10.055

[66] Yuan H, Wang W, Xu D, Xu Q, Xie J, Chen X, Zhang T, Xiong C, He Y, Zhang Y, Liu Y, Shen H. Mint: Outdoor testing and ageing of dye-sensitized solar cells for building integrated photovoltaics. Solar Energy 2018; 165:233-239. DOI: https://doi.org/10.1016/j.solener.2018.03.017

[67] Kato N, Takeda Y, Higuchi K, Takeichi A, Sudo E, Tanaka H, Motohiroa T, Sano T, Toyoda T. Mint: Degradation analysis of dye-sensitized solar cell module after long-term stability test under outdoor working condition. Solar Energy Materials & Solar Cells 2009;93:893-897. DOI: https://doi.org/10.1016/j.solmat.2008.10.022

[68] Berginc M, Krašovec UO, Topič M. Mint: Outdoor ageing of the dye-sensitized solar cell under different operation regimes. Solar Energy Materials & Solar Cells 2014;120:491-499. DOI: http://dx.doi.org/10.1016/j.solmat.2013.09.029

[69] Park N-G, Lagemaat J, Frank AJ. Mint: Comparison of dye-sensitized rutile- and anatase-based TiO_2 solar cells. Journal of Physical Chemistry B 2000;104:8989-8994. DOI: https://doi.org/10.1021/jp9943651

[70] Asghar A, Emziane M, Pak HK, Oh SY. Mint: Outdoor testing and degradation of dye-sensitized solar cells in Abu Dhabi. Solar Energy Materials & Solar Cells 2014;128:335-342. DOI: http://dx.doi.org/10.1016/j.solmat.2014.05.048

[71] Matsui H, Okada K, Kitamura T, Tanabe N. Mint: Thermal stability of dye-sensitized solar cells with current collecting grid. Solar Energy Materials & Solar Cells 2009;93:1110-1115. DOI: 10.1016/j.solmat.2009.01.008

[72] Bella F, Griffini G, Gerosa M, Turri S, Bongiovanni R. Mint: Performance and stability improvements for dye-sensitized solar cells in the presence of luminescent coatings. Journal of Power Sources 2015;283:195-203. DOI: http://dx.doi.org/10.1016/j.jpowsour.2015.02.105

[73] Wu J, Xiao Y, Tang Q, Yue G, Lin J, Huang M, Huang Y, Fan L, Lan Z, Yin S, Sato T. Mint: A large-area light-weight dye-sensitized solar cell based on all titanium substrates with an efficiency of 6.69% outdoors. Advanced Materials 2012;24:1884-1888. DOI: 10.1002/adma.201200003

[74] Freitag M, Teuscher J, Saygili Y, Zhang X, Giordano F, Liska P, Hua J, Zakeeruddin SM, Moser JE, Grätzel M, Hagfeldt A. Mint: Dye-sensitized solar cells for efficient power generation under ambient lighting. Nature Photonics 2017;11:372-378. DOI: 10.1038/NPHOTON.2017.60

[75] Mehmood U, Malik MI, Khan AU, Hussein IA, Harrabi K, Al-Ahmed A. Mint: Effect of outdoor temperature on the power-conversion efficiency of newly synthesized organic photosensitizer based dye-sensitized solar cells. Materials Letters 2018;220:222-225. DOI: https://doi.org/10.1016/j.matlet.2018.03.055

[76] Jeong WS, Lee JW, Jung S, Yun JH, Park NG. Mint: Evaluation of external quantum efficiency of a 12.35% tandem solar cell comprising dye-sensitized and CIGS solar cells. Solar Energy Materials & Solar Cells 2011;95:3419-3423. DOI: 10.1016/j.solmat.2011.07.038

[77] Kubo W, Sakamoto A, Kitamura T, Wada Y, Yanagida S. Mint: Dye-sensitized solar cells: improvement of spectral response by tandem structure. Journal of Photochemistry and Photobiology A: Chemistry 2004;164:33-39. DOI: 10.1016/j.jphotochem.2004.01.024

[78] Lepikko S, Miettunen K, Poskela A, Tiihonen A, Lund PD. Mint: Testing dye-sensitized solar cells in harsh northern outdoor conditions. Energy Science and Engineering 2018;6:187-200. DOI: 10.1002/ese3.195

[79] Roy P, Kim D, Paramasivam I, Schmuki P. Mint: Improved efficiency of TiO_2 nanotubes in dye sensitized solar cells by decoration with TiO_2 nanoparticles. Electrochemistry Communications 2009;11:1001-1004. DOI: https://doi.org/10.1016/j.elecom.2009.02.049

[80] Xing J, Fang WQ, Li Z, Yang HG. Mint: TiO_2-coated ultrathin SnO_2 nanosheets used as photo-anodes for dye-sensitized solar cells with high efficiency. Industrial & Engineering Chemistry Research 2012;51:4247-4253. DOI: https://doi.org/10.1021/ie2030823

[81] Kumara NTRN, Ekanayake P, Lim A, Liew LYC, Iskandar M, Ming LC, Senadeera GKR. Mint: Layered co-sensitization for enhancement of conversion efficiency of natural dye sensitized solar cells. Journal of Alloys and Compound 2013;581: 186-191. DOI: https://doi.org/10.1016/j.jallcom.2013.07.039

[82] Gupta A, Sahu K, Dhonde M, Murty VVS. Mint: Novel synergistic combination of Cu/S co-doped TiO_2 nanoparticles incorporated as photo-anode in dye sensitized solar cell. Solar Energy 2020;203:296-303. DOI: https://doi.org/10.1016/j.solener.2020.04.043

[83] Patni N, Pillai SG, Sharma P. Mint: Effect of using betalain, anthocyanin and chlorophyll dyes together as a sensitizer on enhancing the efficiency of dye-sensitized solar cell. International

Journal of Energy Research. 2020:1-14.
DOI: 10.1002/er.5752

[84] Wood CJ, Summers GH, Gibson EA.
Mint: Increased photocurrent in a
tandem dye-sensitized solar cell by
modifications in push–pull dye-design.
Chemical Communication. 2015;51:
3915-3918. DOI: 10.1039/c4cc10230d

Bifacial Photovoltaic Technology: Recent Advancements, Simulation and Performance Measurement

*Mohammadreza Aghaei, Marc Korevaar, Pavel Babal
and Hesan Ziar*

Abstract

In this chapter, we introduce the physic principle and applications of bifacial PV technology. We present different bifacial PV cell and module technologies as well as investigate the advantages of using bifacial PV technology in the field. We describe the measurement and modeling of Albedo, which is one of the important factors for the energy yield of bifacial PV technology. For an accurate assessment of the performance ratio of bifacial PV strings, it is necessary to measure the albedo irradiance using an albedometer or the front- and rear-side plane of array (POA) irradiance. We also discuss the advanced techniques for the characterization of bifacial PV modules. By means of simulation, we give insight into what boundary conditions result in new bifacial technology gains and the influence of the mounting position of irradiance sensors. We executed several simulations by varying the sensor positions on the rear side of the PV modules, different places, different albedo numbers, mounting heights, different geographical locations with various tilts, seasons, and weather types. To validate the simulation results, we performed various experiments in the field under different conditions. The results prove that the bifacial gain is highly dependent on the mounting heights of PV modules, tilt angles, weather conditions, latitude, and location.

Keywords: bifacial photovoltaics (PV) technology, bifacial solar cell, bifacial PV module, bifaciality factor, solar radiation, Albedo

1. Introduction

In recent decades, photovoltaics (PV) technology has received more attention and PV installation is being dramatically deployed. The PV capacity has been exceeded from one Tera Watt (TW) in the world, which was an impressive milestone in solar energy sector [1]. A novel development is the advent of bifacial PV modules that enhance energy production by converting incident irradiance on the rear side of the module into electricity.

Bifacial solar photovoltaics (PV) cells as a promising technology convert the photons from albedo and incident irradiance into electricity [2]. The bifacial solar PV cell collects the photons simultaneously from both its front and rear sides, whereas the monofacial solar cells can only convert the incident irradiance on the front side [3–5]. The bifacial PV technology as a novel approach to generating electricity with higher performance was investigated by many research groups over five decades [6, 7]. This technology was first applied to the satellites by Russia [2]. In 1970s, Mexican and Spanish researchers released their progress results of bifacial solar cells [8, 9]. Other Spanish research groups obtained the high efficiency and power gain for bifacial PV cells in 1980s [10–12]. In 2000s, the bifaciality concept was introduced for several applications, namely, noise barrier fences and shades [13]. In recent decades, bifacial PV technology has been more considered commercially. In this regard, various companies, such as Yingli Solar, bSolar, Sanyo, and PVG solutions, have produced the bifacial PV modules [14] with different crystalline silicon (c-Si) bifacial PV cell structures, such as PERC (passivated emitter rear contact), PERL (passivated emitter rear locally diffused), PERT (passivated emitter rear totally diffused), IBC (interdigitated back contact), and HIT (heterojunction with intrinsic thin layer).

Bifacial PV modules aim to improve the energy output of PV systems. This is because of doubling the power output of PV through collecting both direct and albedo radiation. Therefore, the bifacial solar cells increase the power density of PV modules [15]. As a result, the PV module and balance of system (BOS) costs and levelized cost of energy (LCOE) are reduced [16].

To date, several large-scale PV power plants were developed using the bifacial c-Si PV modules. For example, a bifacial PV power plant with a capacity of 1.35 MWp was installed in Hokuto city, Japan. The performance data of this plant demonstrated a 21.9% gain compared to a monofacial PV plant with an identical capacity [17]. **Table 1** summarizes the details of some bifacial PV power plants installed around the world.

The cell working temperature is decreased in bifacial solar cells compared to monofacial ones resulting in maximizing the power output [18, 19]. However, the combination of irradiance effect on both front and rear sides of bifacial PV cells resulting a complexity to characterize both sides simultaneously under standard test conditions (STC 1000 W/m^2, AM 1.5 spectrum, and ambient temperature of 25°C). **Figure 1** depicts a scheme of standard bifacial crystalline silicon solar cells. As illustrated in **Figure 1**, an open metallization grid is printed on the front and rear sides of the bifacial structure to be able to collect the incident irradiance from both sides. In n-type bifacial cells, the back surface field (BSF) is n^+, whereas the p^+ diffused layer serves as the emitter, contrary to the p-type bifacial solar cells. The texturized wafers and passivizing anti-reflective coatings (ARC) are partially covered by screen-printed metallic contacts to achieve the open metallization grid [19].

Location	Module technology	PV Plant capacity (MWp)	Annual energy production (GW h)
San Felipe, Chile	Megacell, n-type BiSoN module (BiSoN cell)	2.48	5.78
Eastern US	Sunpreme, GxB370W (Hybrid Cell Technology—HCT)	12.8	8
Hokuto, Japan	PVG Solutions, EarthON 60 (EarthON cell)	1.35	1.47
Datong City, Shanxi, China	Yingli Solar, n-type PANDA module (n-Pasha cell)	50	80
Golmud, China	LONGI Solar	20	—
Golmud, China	Trina Solar	20	—
Golmud, China	Jinzhou Yangguang Energy	20	—
Golmud, China	JA Solar	11	—

Table 1.
The list of bifacial PV power plants using crystalline silicon (c-Si) bifacial modules [8].

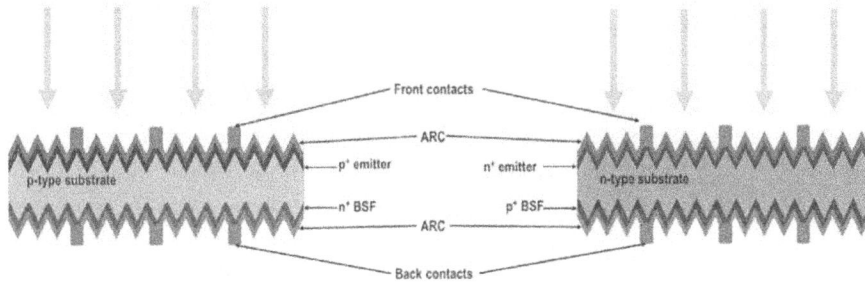

Figure 1.
Standard n-type and p-type bifacial crystalline silicon solar cells [2].

Table 2 summarizes different c-Si bifacial PV cell technologies, including the cross-sectional schematics of each bifacial PV cell structure as well as the efficiency and bifaciality of some cell technologies.

c-Si cell technology	Description	Schematic of the cell structure
Bifacial PERC	• Acronym for passivated emitter rear contact • Efficiency: 19.4–21.2% (front), 16.7–18.1% (rear) [20, 21] • Bifaciality: 80%[20] • Mainly based on p-type crystalline silicon (c-Si) wafer • The cell structure was used for studying the Shockley–Read–Hall recombination velocity at the Si–SiO₂ interface [22] between 1988 and 1991 • It was introduced as a new bifacial PV cell in 1996 [21] • ISF Hamelin [20] and SolarWorld [23] demonstrated the industrial bifacial PERC (or PERC+) design for mass production	
PERL	• Acronym for passivated emitter rear locally diffused • Efficiency: 19.8% (front) [24] • Bifaciality: ≥89% [24] • Mainly based on p-type c-Si wafer • Boron is locally diffused into contact areas at the rear side [25]	
PERT	• Acronym for passivated emitter rear totally diffused • Efficiency: 19.5–22% (front), 17–19% (rear) [26–28] • Bifaciality: ≥85% [26, 29] • Commercialized based on the n-type c-Si wafer; also based on p-type • PVGS developed n-type EarthOn cell in 2009 [29]	

c-Si cell technology	Description	Schematic of the cell structure
	• ECN developed n-Pasha (passivated all-sides H-pattern) cell to reduce yield loss; it was commercialized by Yingli Solar as PANDA [26] • N-MWT (Metal Wrap Through) was also proposed by ECN to narrow down busbars and reduce metal coverage • ISC Konstanz developed BiSoN cell and commercialized by MegaCell	
IBC	• Acronym for interdigitated back contact • Efficiency: 23.2% [30] • Bifaciality: 75% • Mainly based on n-type c-Si wafer • No metal grid contact at the front side, first introduced by Bell Labs in 1954 [31] • ISC Konstanz perfected the production process and introduced ZEBRA cell [32, 33] • ECN proposed Mercury cell with a front floating emitter (FFE) [34, 35]	 ARC/passivation n+ FSF p+ diffused emitter n-type Cz wafer n++ BSF Rear Ag contact ARC/passivation
HIT	• Acronym for heterojunction with intrinsic thin-layer • Efficiency: 24.7% [36] • Bifaciality: ≥95% [37] • Mainly based on n-type c-Si wafer • Introduced by Sanyo (now Panasonic) in 1997 and entered into the serial production under the brand name HITs [36, 38] • Bifacial HITs cell was introduced by Sanyo (now Panasonic) in 2000 and entered the serial production • Basic technology patent was discontinued and opened to the public in 2010; Meyer Burger adopted the cell Technology [39]	 Grid electrode TCO p-type a-Si n-type Cz wafer i-type a-Si n-type a-Si TCO
DSBCSC	• Acronym for double-sided buried contact solar cell • Efficiency: 22% [40, 41] • Bifaciality: 74% [42] • Plated metal contact is entrenched in the laser-formed grove • Low resistance and low shading losses, due to high	 Plated metal (buried contact) Passivation n+ emitter n++ p-type Cz wafer p++ n+ diffused floating junction Plated metal (buried contact) Passivation

c-Si cell technology	Description	Schematic of the cell structure
	metal aspect ratios and lightly diffused emitter [40]	
Silver	• Developed at the Centre for Sustainable Energy Systems, Australian National University [43] • Efficiency: 19.4% [43] • Narrow grooves are created via conventional micro-machining technique, forming a series of thin silicon strips [44] • Reduces silicon consumption by up to a factor of 10 [44]	
BICONTY	• Acronym for bifacial concentrator n-type • Efficiency: 17.6% (front), 16.7% (rear) [45] • Based on the laminated grid n-type solar cell [45] • Silver free—uses copper wire coated with low-temperature solder, resulting in very low shadow loss and resistance loss [45]	

Table 2.
Summary of c-Si bifacial PV cell technologies [8].

In recent years, the bifacial module technology gets more attention in the PV market and numerous attempts have been devoted to the standardization, niche applications, characterization techniques as well as industrial production and costs. The worldwide growth of c-Si bifacial PV cells and modules is predicted by the International Technology Roadmap for Photovoltaic (ITRPV) and it is expected that the bifacial PV technology (c-Si) will be increased by more than 35% by 2028 [46].

2. Albedo: measurement and modeling

Albedo as an important input for the surface energy balance equation [47], thus affects the surface temperature. The speed of chemical and biological reactions is increased due to heat. To overcome this, soil (as a surface) temperature is used to predict the rate at which processes such as seed germination occur. Furthermore, erosion rate and the water content of the soil depend on the ground albedo [48, 49]. This makes albedo utter importance for different environmental processes (e.g., food production) [49]. The local albedo is an influential factor in the heat-island formation that affects public health, namely, air quality and greenhouse gas concentration [50]. Moreover, the energy usage during warm seasons by occupants is related to the value of local albedo [49]. Furthermore, albedo directly influences the electric energy generated by the PV system. Therefore, knowledge about albedo is significant for accurate yield predictions for PV systems, resulting in minimizing technical risks and cost as well as enhancing service life. However, albedo has received less attention so far due to its complexity and minor share in irradiation stroked on a surface of a monofacial PV module. In this regard, albedo is either neglected [51] or assumed to be a constant value [52] in the modeling of PV systems. However, nowadays there are increasing trends of bifacial PV installations and urban PV integration [53], where albedo's contribution to electrical yield becomes more significant [54, 55]. Therefore, better understanding, precise measurements, and accurate estimation of albedo are of utter importance and should be further considered.

2.1 Albedo

Surface albedo[1] is a unitless variable defined as the ratio of the solar radiation at a certain wavelength range reflected from a particular surface (upwelling) to the solar incident upon it.

$$\alpha = \frac{\Phi_S^{downward}}{\Phi_S^{upward}} \tag{1}$$

where α is surface albedo, Φ_S^{down} and Φ_S^{up} are the incoming global radiant fluxes (W) on the down-facing and up-facing sides of the surface S, respectively. Albedo and reflectance are not the same, although often interchangeably used or even misinterpreted. Surface albedo is an angular and spectral integrated value of the spectral bidirectional reflectance distribution function (BRDF):

$$BRDF_\lambda = \rho(\theta_i, \phi_i, \theta_r, \phi_r, \lambda) = \frac{L(\theta_i, \phi_i, \theta_r, \phi_r, \lambda)}{E(\theta_i, \phi_i, \theta_r, \phi_r, \lambda)} \tag{2}$$

where BRDF is the ratio of observed radiance L to incident irradiance E at a wavelength λ under certain viewing geometry (θ_i and φ_i are, respectively, zenith and azimuth angles at the incident direction; θ_r and φ_r represent zenith and azimuth angles, respectively, at the observing direction). Albedo (as a measure used in energy calculations) is normally shown with the Greek letter α, while reflectance (as a property of a material) function is shown by ρ.

Albedo becomes important in the surface energy balance since radiation contains energy. Therefore, it plays a key role in the regulation of earth's surface energy budget [56]. For a piece of land, surface albedo is highly variable, both spatially and temporally. Variations in land coverage and surface conditions, such as snow [57], vegetation [58], urbanization [59], soil moisture [60], sky condition, and position of the Sun, (such as cloudiness [61]), change surface albedo significantly. Thus, surface albedo not only depends on the surface features but also on almost everything that forms the surrounding of surface. In simple terms, at every instant of time, albedo depends on three key factors, namely, material, light source features, and geometry[2] [62].

2.2 Albedo measurement (in situ and satellite)

Albedometer consists of two back-to-back mounted pyranometers and is used for in situ measurements of local albedo. The upper sensor measures the incident global solar radiation and the sensor at the bottom side measures the solar radiation reflected from the surface (s) below. Dividing the obtained value from the bottom

---

[1] Albedo in Latin means whiteness, introduced by Johann Heinrich Lambert in 1760.
[2] Here means arrangement of the surrounding objects in a three-dimensional space.

sensor by that from the upper one gives the value of the albedo at a certain location and time [63]. **Figure 2a** depicts a real albedometer and its schematic is illustrated in **Figure 2b**. The view factor from the albedometer to the target surface makes an influence on the measurement result; hence, the correct distance from the target surface should be opted. Besides, one instant measurement usually does not give the correct average value of albedo, since albedo changes over time. Therefore, long-term measurements of albedo are required. Moreover, the shadow casted by the albedometer itself can reduce the accuracy of the measurement due to the irradiance impinging on the target surface will be different from the incident irradiance on the top pyranometer. The influence of the albedometer shadow on the measurement accuracy depends on the view factor from the albedometer to its shadow [64]. Typically, it is suggested to run albedo measurements during cloudy days to reduce this effect [62]. However, while assessing albedo for larger areas and broader scales is of interest, satellite data are often processed, namely, MODIS (Moderate Resolution Imaging Spectroradiometer) instrument on EOS Terra satellite is used to remotely sense the land albedo of regions on earth.

Satellite albedo data usually have high resolution (in a range of Km). The satellite scans a region on earth once every one or two days [65]. Since albedo is changed over time, hence by the time that a satellite passes over a region on earth and records the albedo, the recorded value may not represent the mean land albedo of the targeted region. However, repeating this procedure for decades gives vital indications of albedo trends for a specific region on earth.

Satellites observe a combined result of the atmosphere and land surface interactions. Therefore, surface albedo is typically retrieved from multispectral optical data considering the viewing geometries. There are common approaches and algorithms used for estimating land surface albedo from satellite optical data. This comprises three following steps—(i) atmospheric correction to filter out the effect of the atmosphere, (ii) BRDF modeling to account for the reflectance anisotropy effect of the land surfaces, and (iii) narrowband to broadband conversion to obtain broadband albedo from spectral albedos, which is available only at certain satellite measurement channels [66].

2.3 Local Albedo models

Most of the local albedo models have been developed between 60's and 90's. The local albedo models rely on empirical coefficients based on long-term data measurements. By the end of 90's, first albedo observing satellites that were orbiting

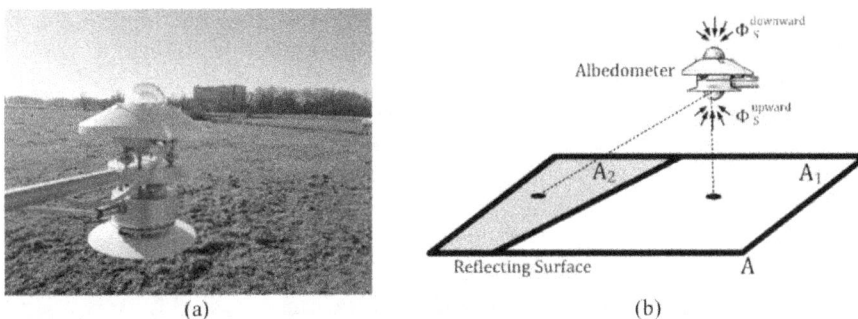

(a)   (b)

Figure 2.
(a) A real albedometer installed on a grass field during landscape survey for photovoltaic installation, and (b) simple schematic of albedometer placement. If the albedo surface A_1 is of interest, surface A_2 will also make an influence on the measurement.

the earth shifted the attention of researchers to the development of algorithms for albedo retrieval from satellite data [67]. However, satellite-derived albedos do not offer enough resolution for complex topography with highly spatial variations, such as in urban areas, where micro-climates and sustainable energies are hot topics [68, 69]. Here, we have reviewed the main local albedo models.

The simplest albedo models are the constant albedo assumption and mean measured albedo, which, respectively, suggest that constant albedo of $\alpha = 0.2$ and $\alpha = \alpha_{site}$ can be applied to all the sites [70, 71]. The constant albedo model may include a considerable error in different situations. The mean measured albedo model demands long-term monitoring of albedo for each site, which is practically impossible.

Another albedo model is the zonal albedo model, which works based on polynomial expressions for the latitude range of $20° < \varphi \leq 60°$ in North America: $\alpha = \sum_{i=1}^{i=3} \alpha_i \varphi^i$, ($\varphi$ is in degrees). The empirical coefficients (α_i) are determined monthly and have been presented in ref. [72]. The drawback of this model is that it cannot be used for local albedo estimation and only is valid for North America. Next model is the *Nkemdirim model* [73], which describes albedo as a function of solar elevation: $\alpha = \alpha_0 \exp (b \cdot \theta_z)$, where θ_z is the solar zenith angle in degrees, and α_0 and b are site-dependent coefficients based on the ground characteristics that should be measured for each location. For this model, the accuracy is dependent on the in situ estimated coefficients. *Beam/diffuse albedo* as a more advanced model separates albedo into its beam (α_b) and diffuse components (α_d), as: $\alpha = f(\alpha_b, \alpha_d)$. This modeling approach requires site-dependent coefficients, which must be determined experimentally for every site [74].

One of the most widely used albedo models is based on the *isotropy assumption*. This model theoretically outputs the albedo using: $\alpha = 0.5 \cdot \alpha_1 (1 - cos\beta)$, where α_1 denotes the albedo coefficient of the site and β is the tilt angle of the surface [75]. This model cannot distinguish the albedo difference for times with equal amounts of global horizontal irradiance and only can consider different values of direct and diffuse components. In this model, the empirical factor of $[1 + sin^2 (Z/2)] (|cos \theta|)]$ is applied as a correction factor for anisotropy of the surface reflection [76], which is called *Temps and Coulson model*.

A more sophisticated model that does not need empirical coefficients has been developed based on the roughness of the surface and the geometry of the surrounding, reflectivity of the materials, and the sky condition. All these parameters are fed into one coherent equation: $\alpha = \sum R_i [C_i \cdot F_{S \to Ai1} + (1/(H+1))(C_i' \cdot F_{S \to Ai1} + F_{S \to Ai2})]$, where R is the spectral reflectivity of the materials forming the surface, $F_{S \to Ai1}$ and $F_{S \to Ai2}$ are, respectively, the view factors from the albedometer to the sunny and shaded parts of the target surface. H is a coefficient dependent on the position of the Sun and the beam and diffuse components of the sunlight. C_i and C_i' are probability-based coefficients calculated by knowing the roughness of the target surface. This model provides better accuracy compared to other local albedo models. This model does not require an empirical coefficient, and also mathematically proves albedo of a surface is always lower than or equal to its reflectivity ($\alpha \leq R$) [62].

3. Indoor characterization of bifacial PV module

For the characterization of bifacial PV modules, the definition and regulation of Standard Test Conditions (STC; 1000 W/m^2 Global irradiance, AM1.5G spectral irradiance, and cell temperature at 25°C) should be extended to consider the spectral and irradiance on rear and front sides of PV modules. In 2019, IEC TS-60904-1-2 was prepared to address the requirements for the characterization of bifacial PV modules [77].

3.1 Single-light source

The single-sided (separate) measurement under STC is a method for indoor characterization of bifacial PV modules, as described by the following equations. In the first step, $BiFi_{I_{sc}}$ is computed according to Eq. (3) [78]:

$$BiFi_{I_{sc}} = \frac{I_{sc,rear}}{I_{sc,front}} \qquad (3)$$

This equation represents the ratio of the short-circuit currents for the front and rear sides of bifacial PV modules. Then $BiFi_{I_{sc}}$ is applied in Eq. (4) to calculate the bifacial equivalent irradiance, G_E:

$$G_E = 1000 Wm^{-2} + BiFi_{I_{sc}} \times G_{rear} \qquad (4)$$

G_E considers the additional contribution of rear-side irradiance on top of the one Sun front irradiance. Subsequently, the front side of the bifacial PV module is flashed with the higher irradiance level of G_E to obtain the bifacial maximum power and efficiency [78]. In this method, the unintended current contribution from the non-illuminated side, which is associated with the reflected irradiance from the surroundings to the optical properties of the module, and to the geometrical disposition of the cells, must be avoided. A common practice involves covering the non-illuminated side of the bifacial module with a black mask [79, 80]. The contribution of unintended illumination could still result in 15% increase in maximum power when using the setup with a black curtain as the background and the rear side of the bifacial PV module covered by a black mask. G. Razongles et al. [81] have derived equations to extrapolate the maximum power rating of bifacial PV modules corresponding to the front- and rear-side illumination. The main equations are given as:

$$P_{max}(S) = S \times f \times P_{max} \qquad (5)$$

$$S = \frac{G_{front}}{1000} \qquad (6)$$

$$f = 1 + a \times (S - 1) + b \times lnS + c \times (S - 1)^2 + d \times (lnS)^2 \qquad (7)$$

Eqs. (4) and (5) were taken from the PV method. The number of Suns, S, refers to the ratio between frontside irradiance and the STC value (1000 W m²). Eq. (7) is the polynomial function of the Neperian logarithm; a, b, c, and d are constants. The parameter f represents the "irradiance coefficient." Pmax is maximum power, which was measured by flashing the front side at an equivalent irradiance, as described in Eq. (8):

$$G_E = G_{front} + G_{back} \times \frac{I_{sc,STC,Back}}{I_{sc,STC,front}} \qquad (8)$$

Alternatively, Corbellini et al. [82] calculated the bifaciality factor using the STC power of the front and rear sides, which is expressed in Eq. (9):

$$BiFi_{Pmax} = \frac{P_{max,rear}}{I_{max,front}} \qquad (9)$$

Accordingly, the unintended irradiance on the non-illuminated side has been quantified that is included in the single-sided power measurements (STC). It was measured repeatedly using a reference cell at several positions on the non-illuminated side. The irradiance ratio on the positions concerning the irradiance on the illuminated side was calculated. This was also measured by a reference cell. The lowest ratio was taken to correct the power measurements (STC) for both the front and rear sides. Moreover, Singh et al. proposed other equations based on the one-diode equivalent model [83] to calculate the power and efficiency of bifacial PV modules.

3.2 A single-light source with reflector or mirrors

Using a reflector that can be located closely behind the rear side of the module, it is possible to obtain the bifaciality of bifacial PV modules. However, a reflective white sheet or background yields irreproducible results. The rear-side illumination condition mostly depends on the reflector properties and light source geometry. Moreover, it depends on the transmittance of the PV cell and module. This probably led to possible difficulties in the quantification of these effects and their standardization [84].

Razongles et al. [81] have used a similar setup (see **Figure 3**) to the Bifacial Cell Tester (BCT) [85]. Soria et al. [86] have conducted bifacial PV mini-module (of four cells) characterization using an identical setup. As shown in **Figure 3**, to reduce the influence of angle, it is possible to optimize the angle between the bifacial PV module and the mirrors (e.g., 44.1° instead of 45°) [86]. Hitherto, this setup has not been scaled up for full-size bifacial PV module characterization.

3.3 Double-light source

For the characterization of full-size bifacial PV modules, new setups with two light sources have been proposed by industrial partners. For instance, Eternal Sun has proposed a setup including two identical units of the solar simulator (see **Figure 4a**) [81]. In this method, a full-size bifacial PV module is sandwiched between both solar simulators for characterization. In another method, an integrated setup (see **Figure 4b**) with two independently controlled xenon flashers (solar simulators) was proposed by h.a.l.m. elektronik [84]. In this technique, the full-size bifacial PV module (under characterization) is secured inside a test

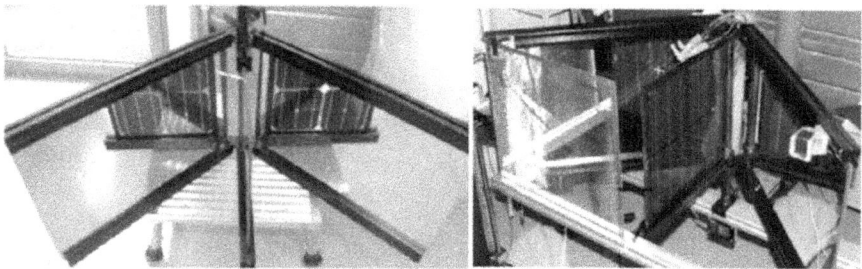

Figure 3.
Two-mirror setup for bifacial PV module characterization [81].

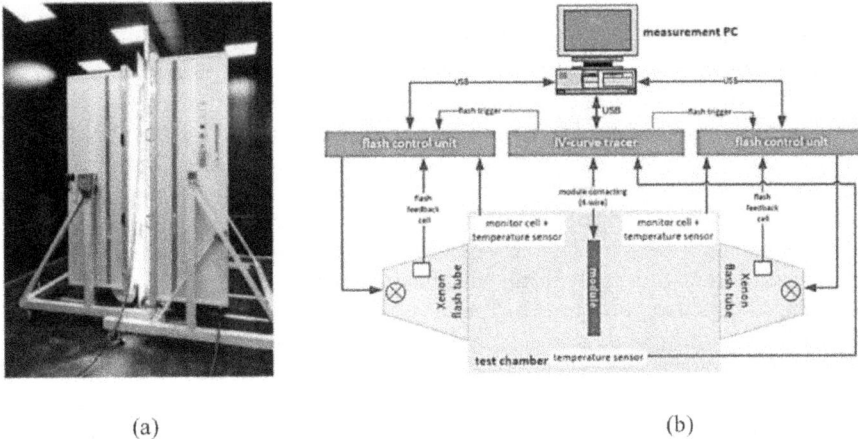

(a)   (b)

Figure 4.
(a) A bifacial illumination setup using two identical solar simulators, proposed by Eternal Sun [81], and (b) A bifacial illumination setup proposed by h.a.l.m elektronik [84].

chamber. Theoretically, both solar simulators are controlled simultaneously by a single control system to avoid systematic error due to lagging or mismatch between the flashing sequence of the simulators. It is also possible that illumination on one side of the bifacial PV module will interfere (constructively or destructively) with the illumination on the other side of the module through transmittance and reflection of irradiance between and through cells, thereby introducing irradiance non-uniformity and inconsistencies. A commercial double-light source solution has been proposed by Swiss company Pasan SA (subsidiary of Meyer-Burger Technology AG) by merely doubling its existing multi-lamp Xenon module simulator. In this method, the distance between the two light sources is approximately 16 meters. This helps to reduce the interferences between them.

However, replacing the xenon lamp with LED light led to enabling variation of spectral irradiances [87, 88]. This aims to enhance the prediction of real-world performance or simulate the rear-side irradiance conditions that are probably closer to the diffuse component of AM1.5G than AM1.5G itself (the standard reference irradiance for both front and rear sides of the module, according to the draft version of IEC TS 60904-1-2). The LED-based solar simulator achieving Class AAA has been made commercially available but is yet to be adapted for simultaneous illumination (double-light sources). In summary, the main fundamental challenges would be related to the spatial requirement and the cost of deploying two solar simulators. Therefore, for justification of the measurement results' accuracy, it is recommended to conduct a cost-benefit analysis for the required setup cost.

4. Simulation of bifacial gain for latitude, tilt, and weather

The bifacial gain defines as an additional amount of power generated by a bifacial PV module over a similar monofacial PV module. The amount of bifacial gain depends on many factors, namely, ground albedo, the height of mounted PV module, the ground coverage ratio, the bifaciality factor of the PV module, and the angular distribution of the incident light. This angular distribution changes with the solar position, influenced by the time of day, season, and location on the earth, as well as by the weather (diffuse cloudy sky as opposed to clear sky). Here, we investigate how big the influence of these factors is on the bifacial gain.

The performance ratio of PV plants with bifacial PV module technology is determined by measuring the plane of array (POA) irradiance and also the irradiance due to ground albedo. Some albedometers (e.g., produced by Kipp & Zonen) can measure either the albedo or the front and rear sides POA, see **Figure 5a**.

Now, the question is that "to what extent does the position of the rear side POA sensor influence the irradiance measurement?" To come to insight into what extent the measurement position influences the irradiance measurement, we have performed a simulation study using "Bifacial Radiance" software [89] on a bifacial PV plant, as sketched in **Figure 5b**. The measurement position has been investigated when it comes to the position on the rear side of fixed PV modules. Additionally, the measurement position for different albedos, locations (and corresponding tilt), the heights of PV module, intra-plant positions as well as month of the year and weather were investigated by several studies [90–92]. The main simulation results of the sensor position are reported in the following sections.

4.1 Simulation method

In this study, the simulations have been executed using "Bifacial Radiance" toolkit that was developed by NREL [89]. This toolkit uses ray-tracing technique to simulate a set of bifacial modules that are either installed on a fixed or a tracking mounting structure. The simulation was run starting at the settings reported in **Table 3**. The general layout is shown in **Figure 5b** to yield the bifacial gain as a function of albedo, height, and location of the site, impacting the tilt angle (here we assumed the tilt angle is equal to the latitude) [93]. The simulation was run to yield the cumulative yearly results. Moreover, the seasonal effect was investigated by selecting certain days and months of the year with different weather conditions. We have imported TMY weather data for the simulation [94].

The simulation was run to yield a distribution of POA^{rear} over the rear of the bifacial PV module and to evaluate optimal positions for measuring irradiance. Subsequently, it was investigated how robust is this value by first varying the position within the PV plant. The position was moved from the edge of the PV module toward the center of the row in a set of steps. Secondly the clearance height, albedo was varied over a large range of values. Later, we assessed the effect of the site location and tilt angle [93].

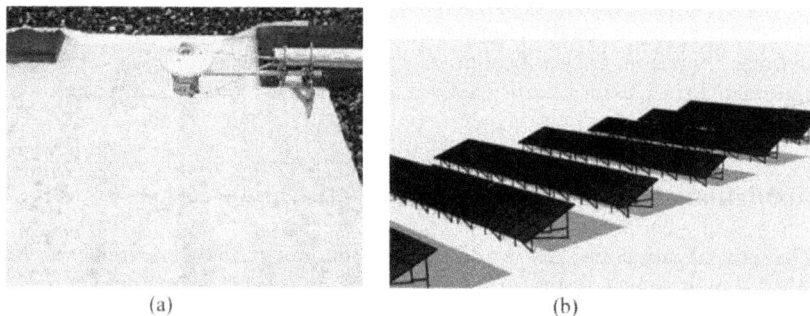

(a)   (b)

Figure 5.
(a) Kipp & Zonen Albedometer mounted above a white surface, and (b) Sketch of bifacial Modules mounted on a fixed mounting structure in seven rows as used in the simulation.

Setting	Value	Setting	Value
Clearance height	0.8 m	Albedo	0.3
No. of sensors	20	Orientation	Landscape
Tilt	25°	Rows	7
Azimuth	180°	No. Modules stacked	4
Plant type	fixed	Modules per row	80
GCR	0.35	Location	Riyadh S.A.

Table 3.
Simulation settings and parameters of the PV plant in bifacial radiance.

4.2 Results

4.2.1 Bifacial gain simulation results

Figure 6a and **b** shows the simulation results of albedo and clearance height. With an increase in albedo, the bifacial gain is linearly increased. For small clearance heights (i.e., modules mounted close to the ground), the bifacial gain also linearly is increased. However, for larger heights, the changes of bifacial gain would be more constant.

Figure 7a and **b** depicts the bifacial gain for a certain time, less than a year for different seasons (1 month) and weather conditions (1 day). The results of bifacial gain for different seasons show that the bifacial gain in summer is significantly higher compared to the bifacial gain in fall or winter.

The weather condition influences the bifacial gain, namely, when the weather is more cloudy then the bifacial gain is increased and vice versa. The seasonal

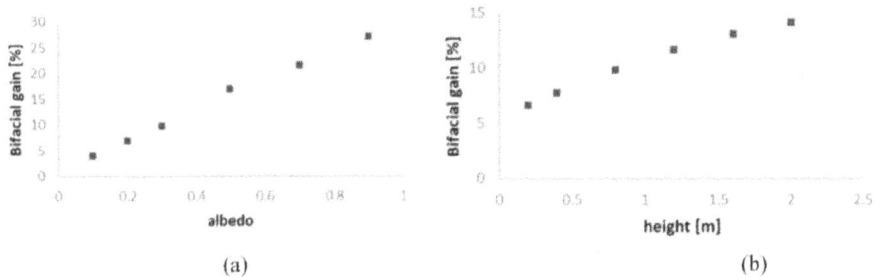

Figure 6.
Bifacial gain as a function of (a) the albedo and (b) the clearance height.

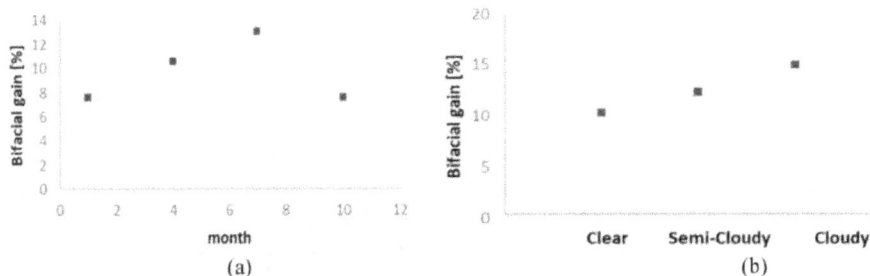

Figure 7.
Bifacial gain as a function of (a) the season and (b) the weather.

conditions also influence the angular distribution of the irradiance incident on the front and rear sides of the bifacial PV modules.

The bifacial gain as a function of location on the earth and the corresponding tilt angle is shown in **Figure 8**. The higher tilt angles of PV module located at a higher latitude resulted in a higher bifacial gain.

4.2.2 Sensor position simulation results

The cumulative yearly average irradiance on the rear side of the PV modules was calculated using "Bifacial Radiance." **Figure 9a** shows the results of located bifacial PV modules in the center of the PV plant using the same input listed in **Table 3**.

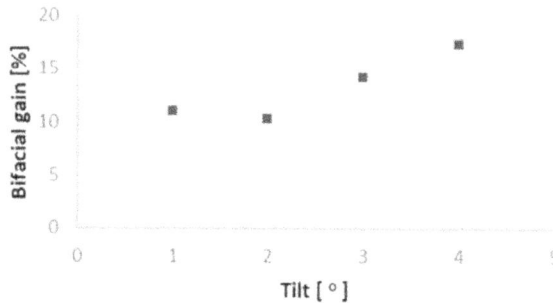

Figure 8.
Bifacial gain as a function of the location with a certain latitude and a tilt angle equal to the latitude.

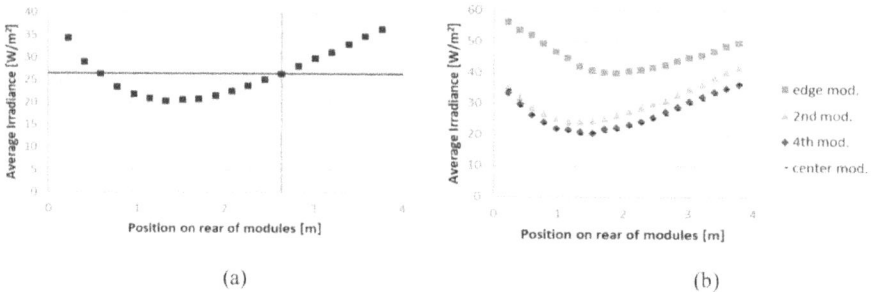

(a)   (b)

Figure 9.
(a) Position dependence of rear side irradiance, where the 0-meter position corresponds to the bottom of 4 modules and the 4-meter position to the top of 4 modules. The red line corresponds to the average incident irradiance on the PV module and the green line corresponds to the 68% position, and (b) Position dependence of rear side irradiance as a function of PV plant position (edge, 2nd, 4th, and center-mounted bifacial PV modules).

The results from selected points in **Figure 9a** are summarized in **Table 4**. As observed in **Table 4**, the pyranometer reading at the 68% position is matched with the average incident irradiance on the PV module. An additional point close to the bottom of the PV module would also be matched, but this point was not chosen since the curve shown in **Figure 9b** is much steeper, which may lead to greater uncertainty.

The results of position effects are shown in **Figure 9b**, where the POA^{rear} was plotted at the edge of a row of PV modules and at a set of positions closer to the center of the row. As shown in **Figure 9b**, at the 4th module in the row there is practically the same amount of irradiance on the rear of the PV module compared to the modules located at the center. **Figure 10** shows the position dependence of rear

Position [%]	$E^{pyranometer}$-$\check{E}$ [%]	Position [%]	$E^{pyranometer}$-$\check{E}$ [%]
3	+ 30	68	0
23	−18	83	+18
48	−19	98	+38

Table 4.
Results of the effect of pyranometer position on measured yearly average POA^rear.

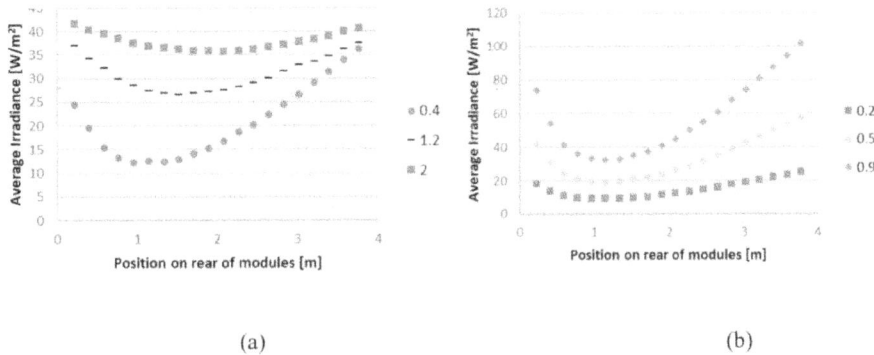

(a)   (b)

Figure 10.
(a) Position dependence of rear-side irradiance as a function of clearance height (0.4, 1.2, and 2), and (b) Position dependence of rear-side irradiance as a function of albedo (0.2, 0.5, and 0.9).

side irradiance as a function of clearance and albedo. As observed in **Figure 10**, the best position for measuring POA^rear irradiance (using a pyranometer) is to be placed sufficiently away from the edge of a row. In this context, when excluding the PV module on the edge side, the shape of POA^rear only is slightly changed for different positions in the PV plant. Therefore, we considered a 68% position given a value corresponding to the irradiance average of the rear side of the bifacial PV module.

The variation of POA^rear irradiance for different clearance heights is shown in **Figure 10a**, from a higher clearance height that led to a larger amount of irradiance on the rear side of the PV module. As illustrated in **Figure 10a**, the POA^rear irradiance is considerably changed when the clearance height is increased from 0.4 meters to 2 meters.

The variation of POA^rear irradiance for different albedos is shown in **Figure 10b**. As expected, a higher albedo led to a greater irradiance on the rear side of the PV module. The POA^rear irradiance does not change noticeably over the range of albedos from 0.2 to 0.9. However, the results show that 68% position gives a reasonable average.

The results for tilt, location, and different positions in the PV plant are shown in **Figure 11**. We have considered different locations, namely, Amsterdam in the Netherlands, Yinchuan in China, Salvador in Brazil, and Riyadh in Saudi Arabia. PV strings have been located in different tilt angles in the selected locations, hence there is a moderate variation in the rear side irradiance magnitude, and also in the shape of the distribution. Therefore, the optimum position of the irradiance sensor should not be strongly dependent on the location of the PV plants.

Figure 12 shows the results for a particular period of the year, certain weather conditions, and different seasons. The results show that the influence of different weather conditions is mostly in the magnitude of the POA^rear irradiance and less in the shape of the irradiance distribution (see **Figure 12a**). For cloudy and semi-cloudy weather conditions, there is more diffuse and less global irradiance in

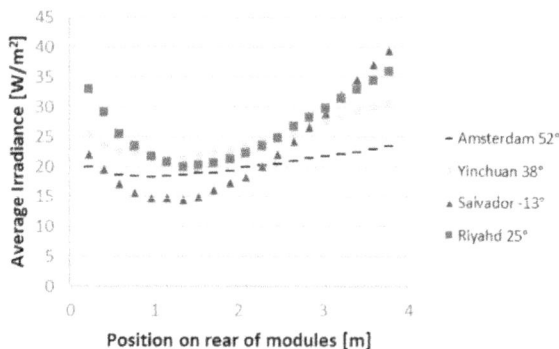

Figure 11.
Position dependence of rear-side irradiance as a function of location and corresponding optimal tilt angles.

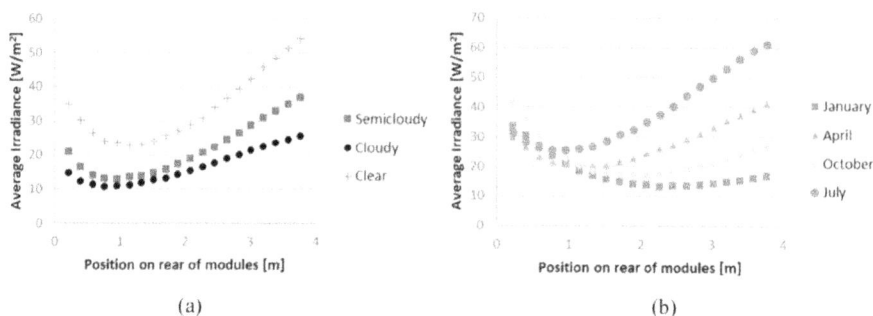

Figure 12.
(a) Position dependence of rear-side irradiance as a function of weather conditions, (b) Position dependence of rear-side irradiance as a function of season.

comparison with a clear sky, which explains the difference in magnitude of POA^{rear} irradiance.

As shown in **Figure 12b**, different seasons significantly influence the shape of POA^{rear} irradiance. For April and July, there is the same amount of irradiance on the rear side of the PV module compared to other ones. However, for higher positions, greater POA^{rear} irradiance on the rear side of the PV module in July and April compared to October and January. This is most likely influenced by the change in solar zenith angle for these months, and to a lesser extent by the change in weather conditions for the months of April and July. Furthermore, when looking at the seasonal data as shown in **Figure 12b**, the irradiance is changed with the season variations. This will have an impact on the optimal mounting position based on the monthly data.

Table 5 presents the results of the difference between measuring at a position of 68% from the rear side and the average irradiance on the PV module for variation of the different parameters. In other words, this would be the average POA^{rear} irradiance difference measured by a pyranometer mounted on the bottom side at the position of 68%. This scores well for changing albedo, height, interplant position, and tilt. For changes in weather and season that are only for a particular period of the year, the 68% position is less representative. This is due to only a particular zenith angle range over a certain (short) period. Strategies for handling this situation can be accepted for a higher measurement uncertainty when looking at the smaller time periods than a year, or alternatively by placing multiple pyranometers at different positions on the rear side of PV modules.

Varied parameter	$|E^{68\%}-\hat{E}|$	Varied parameter	$|E^{68\%}-\hat{E}|$
Interplant position	1.3 %	tilt/location	5.3%
Albedo	2.8 %	Weather	13%
Height	5.6 %	Season	16%

Table 5.
Results of difference between irradiance measured at 68% position and average irradiance for parameter variation.

5. Performance measurement of bifacial PV modules

We performed some experimental measurements for validation of the irradiance distribution on the back of a bifacial PV module at the Kipp & Zonen roof. To be able to choose the optimal sensor(s) and their position(s) for the POA^{rear} irradiance measurement, the simulations need to be compared to data from a real-world measurement setup. This setup needs to provide POA^{rear} sensor data at multiple module height positions as well as the irradiance contribution from the backside of the bifacial module from the total measured module signal. A setup of three rows was constructed where the first row was made of dummy modules for shading, the second middle row contained the modules and sensors, and the third row consisted entirely of modules to simulate proper reflection (as shown in **Figure 13a**). To measure the position-dependent POA^{rear}, there are 6 Kipp & Zonen CMP11 and 6 SPLite 2 pyranometers mounted next to the rear of one of three bifacial modules (**Figure 13b**).

The collected data from the six sensors were compared to simulations done in the Bifacial Radiance software. In **Figure 14**, the simulation model fits well with the measured data from the experimental setup.

The sensor output needs to be compared to the contribution from the POA^{rear} signal of the bifacial PV module. Separating the rear side contribution from the total incident irradiance on the PV module was achieved using different methods, each having various following challenges:

(a)   (b)

Figure 13
(a) Top view of Kipp & Zonen module set up in three rows with two types of bifacial modules above a white albedo surface, and (b) row of six CMP11 pyranometers for position-dependent irradiance measurements on the rear module (red dot in (a)).

1. Covering the front side of the bifacial PV module—module logs only POA^{rear} irradiance.

2. Covering the rear side of one bifacial PV module—module logs only POA irradiance. Subtracting the measured POA irradiance on the first module from the second uncovered module.

3. Using a monofacial PV module with oriented POA^{rear} irradiance.

4. Measuring the POA and POA^{rear} ratio with irradiance sensors and applying it to the PV module output.

It needs to be noted that the PV module output was measured as short-circuit current.

5.1 Bifacial PV module: POArear signal separation methods

5.1.1 Front side of bifacial PV module (POA irradiance) covered

This method allows for an easy cut-out of the POA signal, the realization is observed in **Figure 15a**. The major drawback is that it casts a uniform shadow on the ground without light leaking through the solar cells. The difference between a uniform and representative shadow was measured by placing two irradiance sensors on the ground in the shadow and measuring with the covered and uncovered PV module. The difference is 26–29% less irradiance in case of the uniform shadow. This illustrates the effect of a representative shadow, the influence on the POA^{rear} sensors that would be less than on the ground. This impact is more profound on the measured POA^{rear} compared to the closer setup in the ground (0.5 m tested) and the higher the albedo, as depicted in **Figure 15b**.

5.1.2 Rear side of bifacial PV module (POA^{rear} irradiance) covered

As shown in **Figure 15b**, we have covered the rear side of the bifacial PV module. The idea behind this concept was to measure the POA irradiance from one module by covering its rear side. This POA irradiance data were then subtracted from the adjacent module, giving the POA^{rear} irradiance This method assumes the POA is the same in both modules, although uncertainty may arise from imperfect rear side covering, differences between modules, and slight differences in module mounting angle, which led to acquiring different POA irradiance signal.

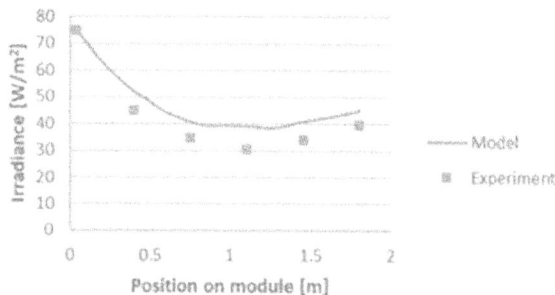

Figure 14.
Measured experimental data from six pyranometers are compared to the simulated model in Bifacial Irradiance software as a function of height along the module.

Figure 15.
(a) Measuring POA^{rear} irradiance with the front side coverage of the PV module , and (b) Measuring POA irradiance with the rear side coverage of the PV module. This is then subtracted from the output of the second uncovered module.

Furthermore, one module is still covered, which casts a uniform shadow. During the day, this shadow extends under another module (from which POA^{rear} will be extracted), reducing the amount of irradiance on its rear side would be absorbed in comparison with the situation of the uncovered PV module.

5.1.3 Monofacial PV module with oriented POA^{rear} irradiance

This method is like the previous method, the advantage being the simplicity of the experiment and set up to get the POA^{rear} irradiance. The monofacial module needs to be a glass-glass module to let light pass through the solar cells to create a representative shadow and similar heating of the module as in case of a bifacial one.

5.1.4 POA and POA^{rear} irradiance sensor ratio applied to the PV module output

The POA and POA^{rear} irradiance are summed up and calculated. This ratio in percentage is applied to the bifacial PV module output. This method suffers the least from the uniform shadow effect. Errors come from the spectral mismatch of the sensors mounted on the front and rear sides of the module.

Among the aforementioned four methods, we have chosen method 2. Method 3 seemed the best but was not used due to the unavailability of monofacial glass-glass modules at the time. The contribution of rear side of the PV module was calculated through Eq. (10).

$$POArearRelative \approx \frac{BifacialPanel - BifacialPanelBackCovered}{BifacialPanel \times \Phi ISC} \qquad (10)$$

where Φ_{ISC} is the module bifaciality factor for short-circuit current. To make a fair comparison with the sensors, both POA and POA^{rear} from each sensor are needed. Each sensor was calibrated on the POA side before being turned around to measure POA^{rear}. Having the relative POA^{rear} contribution from module and sensors, the difference of the sensors to the module POA^{rearRelative} is calculated:

$$POArearRelative = \frac{POArear}{POA + POArear} \qquad (11)$$

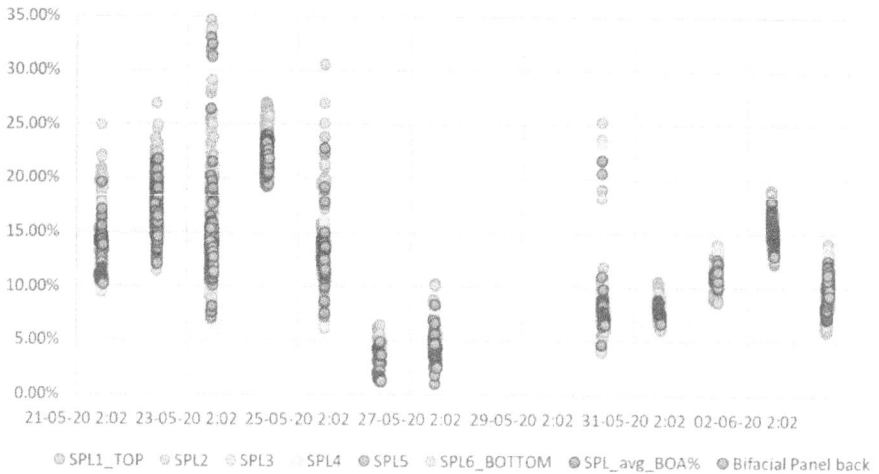

Figure 16.
The comparison of SP Lite2 irradiance sensors and the bifacial module in POA^{rearRelative} under various tilt angles, surface coverages, and weather conditions (only data between 12:00 and 13:30 local time has been considered).

This method is extremely dependent on the accuracy of the POA irradiance sensors. The POA sensors need to be leveled and oriented precisely as same as the orientation of the module. Even more critical, the two PV modules need to be perfectly aligned; otherwise, their relative performance is changed as the sun moves over the day.

In this study, only data from 12:00 to 13:30 local time from each day has been used. This is due to the POA irradiance sensor error mentioned above and due to shadowing from surrounding structures outside of these hours.

5.2 POA^{rear} irradiance on module and sensor comparison

Data were taken from the setup having various module tilt angles (30°, 45°, and 60°) ground covers (white fleece, artificial grass, and roof stones) and experiencing various weather conditions. Generally, the POA^{rear} irradiance contributions are higher during cloudy and overcast weather compared to sunny conditions (see **Figure 16**). This is due to a significant reduction in POA contribution. The highest POA^{rear} contributions are obtained at the lowest angle (30°) and highest albedo using the white fleece (days from 21-5-20 to 25-5-20; see again **Figure 16**). The sensor readings have a higher spread during sunny weather conditions due to the higher direct light contribution, as on the day of 23-5-20 (see again **Figure 16**). The lowest sensor (#6) receives the most irradiance because of higher incident direct light on it compared to other sensors. In cloudy weather conditions, there is only diffuse irradiation, so the spread of the sensor's output is narrow (day 24-5-20; see again **Figure 16**). This is also the case at the highest setup tilt angle (60°) where the effect of direct reflected irradiation is reduced (from day 30-5-20 onwards; see again **Figure 16**). Pyranometers at positions 3 and 4 (60% and 40% of height of module) have the best overlap with the module data.

6. Discussion and conclusions

In recent decades, photovoltaics (PV) technology has received more attention and PV installation is being dramatically deployed. The PV capacity is being approached to one Tera Watt (TWp) in the world. A novel development is the advent of bifacial PV modules that enhance energy production by converting incident irradiance on the rear side of module into electricity. In this chapter, we discussed the physics principle and applications of bifacial PV technology. As discussed, PV plant design considerations influence the bifacial gain. Furthermore, an expected linear relation between the albedo and the bifacial gain was demonstrated in this chapter. Moreover, we proved that mounting the modules at higher heights has a large effect compared to the installed module at a lower height, but once the modules are mounted more than 1.5 meters from the ground, the effect was reduced. To improve the bifacial gain for a particular site, the design of PV strings should be considered by selecting a site with a higher albedo or increasing the albedo artificially by terraforming, for example, using white fleece or stones as ground cover, or regularly moving the vegetation.

We have also discussed that besides the PV plant design, the location and weather conditions play a vital role in the bifacial gain. As shown, at higher latitudes, where corresponding larger tilt angles are used the bifacial gain has been increased. This is most likely caused by the larger tilt of the modules, which ensures that more diffuse irradiance can be absorbed by the rear side of PV modules as well as more soil that scatters direct irradiance from the sun.

In cloudy weather conditions, bifacial gains have been increased, most likely because cloudy conditions tend to have a larger diffuse component in the sky irradiance compared to clear conditions. In addition to that, the direct component in cloudy conditions was decreased thereby significantly reducing the front side irradiance on the PV module. Therefore, it can be said that in the locations with higher tilt angles as well as more cloudy weather, the position percentage is increased in power output by installing bifacial modules instead of monofacial ones that are greater than for lower tilt angles and clear weather.

However, as clear weather and lower tilt angles (latitudes) promote a higher absolute energy generation, the kilowatt-hour (kWh) gain might be larger in clear weather and lower latitudes as opposed to larger latitudes or cloudy conditions.

It should also be mentioned that there is a seasonal component to the bifacial gain, namely, in the summer months there is a larger bifacial gain than in the winter months. This is most likely due to the increasing irradiance of the land close to the modules with higher sun angles as it is typical in the summertime. In this way, the POA^{rear} irradiance of the module is increased as more light scatters from the surface to the rear side of the bifacial PV module.

In summary, in this chapter, we also presented the advantages and performance measurements of bifacial PV technologies. We discussed the recent characterization techniques of bifacial PV modules. Furthermore, for accurately knowing the performance ratio of PV strings with bifacial modules, we explained how to measure the irradiance due to ground albedo. This can be performed with either an albedometer or the front and rear-side plane of array (POA) irradiance very accurately.

By means of simulation using Bifacial Radiance software developed by NREL, we investigated insight into what boundary conditions result in new bifacial technology gains and the influence of the mounting position of irradiance sensors. To this end, we performed different experiments by varying the sensor positions on the rear side of the PV modules in different locations within the PV plant, different albedo numbers, mounting heights, as well as different geographical locations with different tilts, different seasons, and weather types, are investigated. Validation of the results of the simulation using multiple sensors in the Kipp & Zonen roof setup was presented.

Simulation results demonstrated that the irradiance on the rear side of the PV modules varies strongly with its positions. The results also proved that the rear side irradiance distribution and magnitude were varied with albedo, clearance height, and tilt. For the plant of four landscape-positioned modules above each other, an installation position on the rear side of the module at approximately 68% represented the average irradiance well in case of cumulative yearly results. We recommend that for other designs a similar optimal position for yearly results can be found but at a slightly different position.

According to the results, we advise to install the POA^{rear} sensor (s) significantly away from the start or end of a string. Seasonal results or results for a day with weather showed significant changes in the rear side irradiance pattern, impacting the optimal sensor position for seasonal analysis. A possible mitigation strategy is to use multiple pyranometers at different positions.

Author details

Mohammadreza Aghaei[1*], Marc Korevaar[2,3], Pavel Babal[2,3] and Hesan Ziar[4]

1 Department of Ocean Operations and Civil Engineering, Norwegian University of Science and Technology (NTNU), Alesund, Norway

2 Kipp and Zonen, Delft, The Netherlands

3 OTT Hydromet B.V., Kempten, Germany

4 Delft University of Technology (TU Delft), Photovoltaic Materials and Devices, Delft, Netherlands

*Address all correspondence to: mohammadreza.aghaei@ntnu.no

IntechOpen

References

[1] Available from: https://www.pv-magazine.com/2022/03/15/humans-have-installed-1-terawatt-of-solar-capacity/ (n.d.)

[2] Guerrero-Lemus R, Vega R, Kim T, Kimm A, Shephard LE. Bifacial solar photovoltaics – A technology review. Renewable and Sustainable Energy Reviews. 2016;**60**:1533-1549

[3] Aghaei M, Nitti M, Ekins-Daukes NJ, Reinders AHME. Simulation of a novel configuration for luminescent solar concentrator photovoltaic devices using bifacial silicon solar cells. Applied Sciences. 2020;**10**(3):1-10. DOI: 10.3390/app10030871

[4] Aghaei M, Pelosi R, Wong WWH, Schmidt T, Debije MG, Reinders AHME. Measured power conversion efficiencies of bifacial luminescent solar concentrator photovoltaic devices of the mosaic series. Progress in Photovoltaics Research and Applications. 2022:1-14. DOI: 10.1002/pip.354614AGHAEIET AL

[5] Aghaei M, Zhu X, Debije MG, Wong WWH, Schmidt T, Reinders AHME. Simulations of Luminescent Solar Concentrator Bifacial Photovoltaic Mosaic Devices Containing Four Different Organic Luminophores. IEEE Journal of Photovoltaics. 2022;**3**(3): 771-777. DOI: 10.1109/JPHOTOV. 2022.3144962

[6] Luque A, Cuevas A, Ruiz J. Double-sided n+-p-n+ solar cell for bifacial concentration. Solid Cells. 1980;**2**(2): 151-166

[7] Hiroshi M. Radiation energy transducing device. In: Google Patents. 1966

[8] Liang TS et al. A review of crystalline silicon bifacial photovoltaic performance characterisation and simulation. Energy & Environmental Science. 2019;**12**:116-148. DOI: 10.1039/c8ee02184h

[9] Chambouleyron I, Chevalier Y. Silicon double solar cell. In: Photovoltaic Solar Energy Conference. 1978. pp. 967-976

[10] Cuevas A, Luque A, Eguren J, Del Alamo J. High efficiency bifacial back surface field solar cells. Solid Cells. 1981;**3**(4):337-340

[11] Cuevas A, Luque A, Eguren J, del Alamo J. 50 Per cent more output power from an albedo-collecting flat panel using bifacial solar cells. Solar Energy. 1982;**29**(5):419-420

[12] Luque A, Lorenzo E, Sala G, López-Romero S. Diffusing reflectors for bifacial photovoltaic panels. Solid cells. 1985;**13**(3):277-292

[13] Hezel R. Novel applications of bifacial solar cells. Progress in Photovoltaics: Research and Applications. 2003;**11**(8):549-556

[14] Halm A et al. Encapsulation losses for ribbon contacted N-type IBC solar cells. In: 29th European Photovoltaic Solar Energy Conference and Exhibition (EU PVSEC 2014). 2014. pp. 190-193. DOI: 10.4229/EUPVSEC20142014-1BV.6.49

[15] Kreinin L, Bordin N, Karsenty A, Drori A, Eisenberg N. Experimental analysis of the increases in energy generation of bifacial over mono-facial PV modules. In: Proceedings of the 26th Europe Photovoltaic Solar Energy Conference Exhibition. Hambourg, Genmany; 2011. pp. 5-9

[16] Hauser A, Richter A, Leu S. Cell and Module Design from the LCOE Perspective. Semantic Scholar. Meyer Burg; 2014. https://www.semanticsch

olar.org/paper/Cell-and-module-design-from-the-LCOE-perspectiveHauser/af0969f73251b9424a91bcd5c17c42a164e2a200

[17] Chiodetti M. Bifacial PV plants: Performance model development and optimization of their configuration. Sweden: KTH Royal Institute of Technology; 2015

[18] Sugibuchi K, Ishikawa N, Obara S. Bifacial-PV power output gain in the field test using 'EarthON' high bifaciality solar cells. In: Proc. 28th PVSEC. 2013. pp. 4312-4317

[19] Yang L et al. High efficiency screen printed bifacial solar cells on monocrystalline CZ silicon. Progress in Photovoltaics: Research and Applications. 2011;**19**(3):275-279

[20] Dullweber T et al. PERC+: Industrial PERC solar cells with rear Al grid enabling bifaciality and reduced Al paste consumption. Progress in Photovoltaics: Research and Applications. 2016;**24**(12):1487-1498

[21] Hübner A, Aberle AG, Hezel R. Novel cost-effective bifacial silicon solar cells with 19.4% front and 18.1% rear efficiency. Applied Physics Letters. 1997;**70**(8):1008-1010

[22] Aberle AG, Glunz S, Warta W. Impact of illumination level and oxide parameters on Shockley–Read–Hall recombination at the Si-SiO$_2$ interface. Journal of Applied Physics. 1992;**71**(9):4422-4431

[23] Solar World. Calculating the additional energy yield of bifacial solar modules. 2016. Available from: https://solaren-power.com/pdf/Calculating-Additional-Energy-Yield-Through-Bifacial-Solar-Technology.pdf

[24] Lohmüller E et al. Bifacial p-type silicon PERL solar cells with screen-printed pure silver metallization and

89% bifaciality. In: 33rd EU PVSEC. 2017. pp. 418-423

[25] Green MA, Blakers AW, Zhao J, Milne AM, Wang A, Dai X. Characterization of 23-percent efficient silicon solar cells. IEEE Transactions on Electron Devices. 1990;**37**(2):331-336

[26] Tool CJJ et al. Industrial cost effective N-PASHA solar cells with ampersand >20% efficiency. ECN. 2013

[27] Cai W et al. 22.2% efficiency n-type PERT solar cell. Energy Procedia. 2016;**92**:399-403

[28] Wei Q, Zhang S, Yu S, Lu J, Lian W, Ni Z. High efficiency n-PERT solar cells by B/P co-diffusion method. Energy Procedia. 2017;**124**:700-705

[29] Ishikawa N. Industrial production of high efficiency and high bifaciality solar cells. 2012. [Online]. Available: https://pvpmc.sandia.gov/resources-and-events/events/2012-bifacial-pv-workshop-konstanz/

[30] Galbiati G, Chu H, Mihailetchi VD, Libal J, Kopecek R. Latest results in screen-printed IBC-ZEBRA solar cells. In: 2018 IEEE 7th World Conference on Photovoltaic Energy Conversion (WCPEC)(A Joint Conference of 45th IEEE PVSC, 28th PVSEC & 34th EU PVSEC). 2018. pp. 1540-1543

[31] Kopecek R et al. Bifaciality: One small step for technology, one giant leap for kWh cost reduction. Photovoltaics Institute. 2014;**26**:32-45

[32] Halm A et al. The zebra cell concept-large area n-type interdigitated back contact solar cells and one-cell modules fabricated using standard industrial processing equipment. In: Proc. the 27th EU Photovoltaic Specialists Conference. 2012. pp. 567-570

[33] Galbiati G et al. "Large-area back-contact back-junction solar cell with

efficiency exceeding 21%," in 2012 IEEE 38th Photovoltaic Specialists Conference (PVSC) PART 2. 2012. pp. 1–6.

[34] Cesar I et al. Enablers for IBC: Integral cell and module development and implementation in PV industry. Energy Procedia. 2017;**124**:834-841

[35] Mewe A et al. Mercury: Industrial IBC cell with front floating emitter for 20.9% and higher efficiency. In: 2015 IEEE 42nd Photovoltaic Specialist Conference (PVSC). 2015. pp. 1-6

[36] Taguchi M et al. 24.7% record efficiency HIT solar cell on thin silicon wafer. IEEE Journal of Photovoltaics. 2013;**4**(1):96-99

[37] Janssen GJM et al. Minimizing the polarization-type potential-induced degradation in PV modules by modification of the dielectric antireflection and passivation stack. IEEE Journal of Photovoltaics. 2019; **9**(3):608-614. DOI: 10.1109/ JPHOTOV.2019.2896944

[38] Kinoshita EMT, Fujishima D, Yano A, Ogane A, Tohoda S, Matsuyama K, et al. The approaches for high efficiency HITTM solar cell with very thin (<100 μm) silicon wafer over 23% 2011. pp. 871-874

[39] Paper W. Heterojunction Technology. 2015. Available from: https://www.meyerburger.com/sg/en/ meyer-burger/media/multimedia/ publications/article/heterojunction-technology-the-solar-cell-of-the-future/

[40] Ghozati SB, Ebong AU, Honsberg CB, Wenham SR. Improved fill-factor for the double-sided buried-contact bifacial silicon solar cell. Solar Energy Materials & Solar Cells. 1998;**51**(2):121-128

[41] Wenham SR, Chan BO, Honsberg CB, Green MA. Beneficial and constraining effects of laser scribing in buried-contact solar cells. Progress in

Photovoltaics: Research and Applications. 1997;**5**(2):131-137

[42] Ebong AU, Lee SH, Honsberg C, Wenham SR. High efficiency double sided buried contact silicon solar cells. Japanese Journal of Applied Physics. 1996;**35**(4R):2077

[43] Verlinden PJ et al. Sliver® solar cells: A new thin-crystalline silicon photovoltaic technology. Solar Energy Materials & Solar Cells. 2006; **90**(18–19):3422-3430

[44] Weber K, Everett V, Franklin E, Blakers A. Results of a Cost Model for Sliver Cells. 2006

[45] Untila GG et al. Bifacial concentrator Ag-free crystalline n-type Si solar cell. Progress in Photovoltaics: Research and Applications. 2015;**23**(5):600-610

[46] Metz A, Fischer M, Trube J. "Recent results of the international technology roadmap for photovoltaics (ITRPV)." ITRPV, 2017

[47] Bristow KL. On solving the surface energy balance equation for surface temperature. Agricultural and Forest Meteorology. 1987;**39**(1):49-54

[48] Idso SB, Jackson RD, Reginato RJ, Kimball BA, Nakayama FS. The dependence of bare soil albedo on soil water content. Journal of Applied Meteorology and Climatology. 1975; **14**(1):109-113

[49] Matthias AD et al. Surface roughness effects on soil albedo. Soil Science Society of America Journal. 2000;**64**(3):1035-1041

[50] Taha H. Urban climates and heat islands: Albedo, evapotranspiration, and anthropogenic heat. Energy and Buildings. 1997;**25**(2):99-103

[51] Kirn B, Brecl K, Topic M. A new PV module performance model based on

separation of diffuse and direct light. Solar Energy. 2015;**113**:212-220

[52] Calcabrini A, Ziar H, Isabella O, Zeman M. A simplified skyline-based method for estimating the annual solar energy potential in urban environments. Nature Energy. 2019;**4**(3):206-215

[53] Wood Mackenzie. Global bifacial module market report 2019. 2020. [Online]. Available: https://www.woodmac.com/reports/power-markets-global-bifacial-module-market-report-2019-348173

[54] Brennan MP, Abramase AL, Andrews RW, Pearce JM. Effects of spectral albedo on solar photovoltaic devices. Solar Energy Materials & Solar Cells. 2014;**124**:111-116

[55] Russell TCR, Saive R, Augusto A, Bowden SG, Atwater HA. The influence of spectral albedo on bifacial solar cells: A theoretical and experimental study. IEEE Journal of Photovoltaics. 2017;**7**(6):1611-1618

[56] Liang S, Zhang X, He T, Cheng J, Wang D. Remote sensing of the land surface radiation budget. Remote Sens. energy fluxes soil moisture content. 2013:121-162

[57] He T, Liang S, Yu Y, Wang D, Gao F, Liu Q. Greenland surface albedo changes in July 1981–2012 from satellite observations. Environmental Research Letters. 2013;**8**(4):44043

[58] Loarie SR, Lobell DB, Asner GP, Mu Q, Field CB. Direct impacts on local climate of sugar-cane expansion in Brazil. Nature Climate Change. 2011;**1**(2):105-109

[59] Offerle B, Jonsson P, Eliasson I, Grimmond CSB. Urban modification of the surface energy balance in the West African Sahel: Ouagadougou, Burkina Faso. Journal of Climate. 2005;**18**(19):3983-3995

[60] Govaerts Y, Lattanzio A. Estimation of surface albedo increase during the eighties Sahel drought from Meteosat observations. Global and Planetary Change. 2008;**64**(3–4):139-145

[61] Seneviratne SI et al. Land radiative management as contributor to regional-scale climate adaptation and mitigation. Nature Geoscience. 2018;**11**(2):88-96

[62] Ziar H, Sönmez FF, Isabella O, Zeman M. A comprehensive albedo model for solar energy applications: Geometric spectral albedo. Applied Energy. 2019;**255**:113867

[63] Kipp & Zonnen BV. Albedometers. 2015

[64] Sönmez FF, Ziar H, Isabella O, Zeman M. Fast and accurate ray-casting-based view factor estimation method for complex geometries. Solar Energy Materials & Solar Cells. 2019;**200**:109934

[65] NASA. Moderate Resolution Imaging Spectrometer. 2022. https://modis.gsfc.nasa.gov

[66] He T et al. Evaluating land surface albedo estimation from Landsat MSS, TM, ETM+, and OLI data based on the unified direct estimation approach. Remote Sensing of Environment. 2018;**204**:181-196

[67] Lucht W, Schaaf CB, Strahler AH. An algorithm for the retrieval of albedo from space using semiempirical BRDF models. IEEE Transactions on Geoscience and Remote Sensing. 2000;**38**(2):977-998

[68] Enriquez R, Zarzalejo L, Jiménez MJ, Heras MR. Ground reflectance estimation by means of horizontal and vertical radiation measurements. Solar Energy. 2012;**86**(11):3216-3226

[69] Psiloglou BE, Kambezidis HD. Estimation of the ground albedo for the Athens area, Greece. Journal of Atmospheric and Solar—Terrestrial Physics. 2009;**71**(8–9):943-954

[70] Liu BYH, Jordan RC. The long-term average performance of flat-plate solar-energy collectors: With design data for the US, its outlying possessions and Canada. Solar Energy. 1963;**7**(2):53-74

[71] Ineichen P, Perez R, Seals R. The importance of correct albedo determination for adequately modeling energy received by tilted surfaces. Solar Energy. 1987;**39**(4):301-305

[72] Gueymard C. Mathermatically integrable parameterization of clear-sky beam and global irradiances and its use in daily irradiation applications. Solar Energy. 1993;**50**(5):385-397

[73] Nkemdirim LC. A note on the albedo of surfaces. Journal of Applied Meteorology. 1972;**11**(5):867-874

[74] Ineichen P, Guisan O, Perez R. Ground-reflected radiation and albedo. Solar Energy. 1990;**44**(4):207-214

[75] Hay JE. Calculating solar radiation for inclined surfaces: Practical approaches. Renewable Energy. 1993;**3** (4–5):373-380

[76] Temps RC, Coulson KL. Solar radiation incident upon slopes of different orientations. Solar Energy. 1977;**19**(2):179-184

[77] IEC. Photovoltaic Devices–Part 1-2: Measurement of Current-Voltage Characteristics of Bifacial Photovoltaic (PV) Devices; IEC TS 60904-1-2:2019, IEC TS. 2019

[78] Deline C, MacAlpine S, Marion B, Toor F, Asgharzadeh A, Stein JS. Assessment of bifacial photovoltaic module power rating methodologies—

inside and out. IEEE Journal of Photovoltaics. 2017;**7**(2):575-580

[79] Nussbaumer H et al. Accuracy of simulated data for bifacial systems with varying tilt angles and share of diffuse radiation. Solar Energy. 2020; **197**:6-21

[80] Singh JP, Guo S, Peters IM, Aberle AG, Walsh TM. Comparison of glass/glass and glass/backsheet PV modules using bifacial silicon solar cells. IEEE Journal of Photovoltaics. 2015;**5**(3): 783-791

[81] Razongles G et al. Bifacial photovoltaic modules: Measurement challenges. Energy Procedia. 2016;**92**: 188-198

[82] Corbellini G, Vasco M. Analysis and modelling of bifacial PV modules. In: 14th PV Performance Modeling Collaborative Workshop Modelling of Bifacial PV Modules. 2015

[83] Singh JP, Aberle AG, Walsh TM. Electrical characterization method for bifacial photovoltaic modules. Solar Energy Materials & Solar Cells. 2014; **127**:136-142

[84] Edler A, Metz A. Need for standardized measurements of bifacial solar cells and PV modules. In: 14th PV Performance Modeling Collaborative Workshop Modelling of Bifacial PV Modules. 2015

[85] Ezquer M, Petrina I, Cuadra JM, Lagunas A, Cener-Ciemat F. Design of a special set-up for the IV characterization of bifacial photovoltaic solar cells. 2009

[86] Soria B, Gerritsen E, Lefillastre P, Broquin JE. A study of the annual performance of bifacial photovoltaic modules in the case of vertical facade integration. Energy Science Engineering. 2016;**4**:52-68. DOI: 10.1002/ese3.103

[87] Bliss M, Betts TR, Gottschalg R. Indoor measurement of photovoltaic device characteristics at varying irradiance, temperature and spectrum for energy rating. Measurement Science and Technology. 2010;**21**(11):115701

[88] Stuckelberger M et al. Class AAA LED-based solar simulator for steady-state measurements and light soaking. IEEE Journal of Photovoltaics. 2014;**4** (5):1282-1287

[89] Deline C, Marion W, Ayala S. Bifacial Radiance. 2017. Available from: https://github.com/NREL/bifacial_ radiance

[90] Riedel-Lyngskær N, Petit M, Berrian D, Poulsen PB, Libal J, Jakobsen ML. A spatial irradiance map measured on the rear side of a utility-scale horizontal single axis tracker with validation using open source tools. In: 2020 47th IEEE Photovoltaic Specialists Conference (PVSC). 2020. pp. 1026-1032

[91] Pelaez SA, Deline C, Greenberg P, Stein JS, Kostuk RK. Model and validation of single-axis tracking with bifacial PV. IEEE Journal of Photovoltaics. 2019;**9**(3):715-721

[92] Pelaez SA, Deline C, MacAlpine SM, Marion B, Stein JS, Kostuk RK. Comparison of bifacial solar irradiance model predictions with field validation. IEEE Journal of Photovoltaics. 2018; **9**(1):82-88

[93] Landau CR. Optimum tilt of solar panels. 2014. Available from: https:// www.solarpaneltilt.com/

[94] NREL. Weather Data. Available from: https://energyplus.net/weather [Accessed: 20.01.2020]

Chapter 9

Photovoltaic Power Forecasting Methods

Ismail Kaaya and Julián Ascencio-Vásquez

Abstract

The rapid growth in grid penetration of photovoltaic (PV) calls for more accurate methods to forecast the performance and reliability of PV. Several methods have been proposed to forecast the PV power generation at different temporal horizons. In this chapter the different methods used in PV power forecasting are described with an example on their applications and related uncertainty. The methods discussed include physical, heuristic, statistical and machine learning methods. When benchmarked, it is shown that physical method showed the highest uncertainties compared to other methods. In the chapter, the effect of degradation on lifetime PV power and energy forecast is also assessed using linear and non-linear degradation scenarios. It is shown that the relative difference in lifetime yield prediction is over 5% between linear and non-linear scenarios.

Keywords: Degradation, Lifetime, Photovoltaic, Power, Forecasting

1. Introduction

The current trends in photovoltaic (PV) deployments worldwide are a clear indication that PV energy will play a big role in the near future energy mix. For example, the global solar photovoltaic (PV) capacity is projected to increase by 37.5% from 2019 to 2030 (i.e from 593.9 GW in 2019 to 1,582.9 in 2030) [1]. This rapid growth in grid penetration of PV calls for more accurate methods to forecast the performance and reliability of PV. Additionally, PV projects policy and investment decisions rely on PV performance and reliability forecasts. Therefore, to reduce the risks of PV investments, more reliable methods to forecast the performance and reliability of PV power are a prerequisite.

Different methods have been proposed for PV power forecasting. These methods can be classified as: physical, heuristic, statistical and machine learning methods [2, 3]. Each method might have different conceptual design, implementation, application and accuracy. In this chapter, the application and accuracy of the different methods are assessed using measured PV module power and weather data.

PV power forecasting can range from different temporal horizons depending on ones need. At a moment there is no standard classification criterion of the temporal horizon. General classification can be made as: very short term (Intra-hour: 15 minutes to 2 hours ahead), short-term (hour ahead: 1 to 6 hours ahead, day ahead: 1–3 days ahead), Medium-term (week to months ahead), long-term (one to several years) and lifetime forecast (until PV expected lifetime).

Achieving high-accuracy forecasts at each of these temporal horizons is influenced by different variables such as: solar radiation models, PV performance

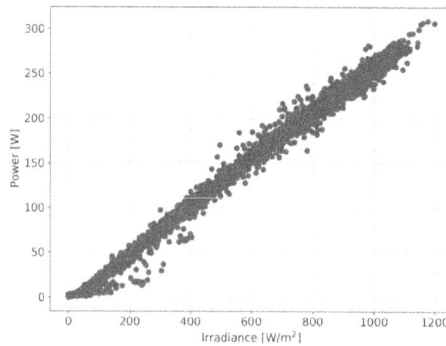

Figure 1.
Measured irradiance versus measured power. Data corresponds to six month measurements of irradiance and module power in Gran Canaria (Spain).

models, data availability, data quality and forecasting methods. The accuracy highly deteriorates with increasing forecasting temporal horizon. This is because, more input parameters such as: seasonal variations, soiling effects, degradation and many other performance reducing effects need to be taken into consideration [4]. These factors are not easy to evaluate precisely, therefore, they are simply approximated which increase the uncertainties in long-term and lifetime PV forecast.

The main influencing factor of PV production is the amount of global solar irradiation incident on the PV panels. As shown in **Figure 1**, there is a quasi-linear correlation of power and irradiance. This property means that the accuracy of power prediction is highly determined by the accuracy of the solar irradiation forecast.

In this chapter, the different methods used in PV power forecasting are presented. The chapter is organised as follows: In Section 1, a brief introduction on power forecasting is presented. In Section 2, different PV forecasting methods are presented, for some methods a practical example of their application and their accuracy are evaluated using real measured PV module power. Section 3, is dedicated to lifetime PV power forecasting. In this section, several effects affecting lifetime PV power forecasting are stated and a more elaborative discussion of the degradation effect on lifetime PV power forecasting is presented. Lastly, in Section 4, a summary of the different aspects within the chapter is presented.

2. PV power forecasting methods

Different methods: physical, heuristic, statistical and machine learning are commonly used in PV power forecasting [2, 3]. The methods are based on two main approaches to generate the PV power forecasting. One is the physical approach, which requires prior knowledge of PV material properties and the metadata of a PV system, together with the need of weather data. The second ones is the data-driven approach, which requires operational data to train/calibrate coefficients of the models which are then used to generate the predictions. This means that, a data-driven approach can only be applied after a given PV module or system has been exposed and enough data is available to train/calibrate the models. On-contrary, a physical approach can be applied even when the PV system is not yet commissioned. Hence PV power forecasting methods based on a physical approach are the mostly used methods by PV stakeholders to evaluate the economic viability of PV projects during the initial phases.

Figure 2 illustrates the required inputs and general steps to generate PV power forecast by the two approaches. What is clear is that both approaches require weather data (mainly solar irradiation) as input. Therefore, solar irradiation forecast is highly essential step for PV power forecast using both approaches. Unlike data-driven approach, physical approach is based on physical assumptions and therefore, knowledge of the physical parameters influencing PV generation is required.

2.1 Physical method

Physical models calculate the PV power using the equivalent electrical circuits [5]. The equivalent circuit model developed for a single cell can be used to derive equivalent circuit models for a PV module as well as a PV system [6]. They are the commonly applied models in the PV power forecasting commercial software packages (such as PVSyst [7] PVWATTS [8]).

2.1.1 PV cell and module equivalent circuit models

To build the physical model one need to know the basic photo-to-voltage theory. The diode model is used to develop the PV cell model to calculate the PV output power. The diode model can be characterized as: one-, two- and three-diode models [5, 7] (see **Figure 3**). The choice of the model selection depend on charge carrier recombination mechanisms one need to take into consideration. Because of its simplest, the one-diode model is the most commonly used to model the operating principles of a PV cell. The one-diode model can consist of three, four or five parameters (see **Table 1**) .

The three parameter (3-P) one-diode model is only used to demonstrate the basic working principles of a PV cell but not as a representative of the real operating condition. To simulate the actual working conditions of a PV cell, the five parameter (5-P) one-diode model is commonly used because it takes into account both the series and shunt resistive losses. It is expressed as [9, 10]:

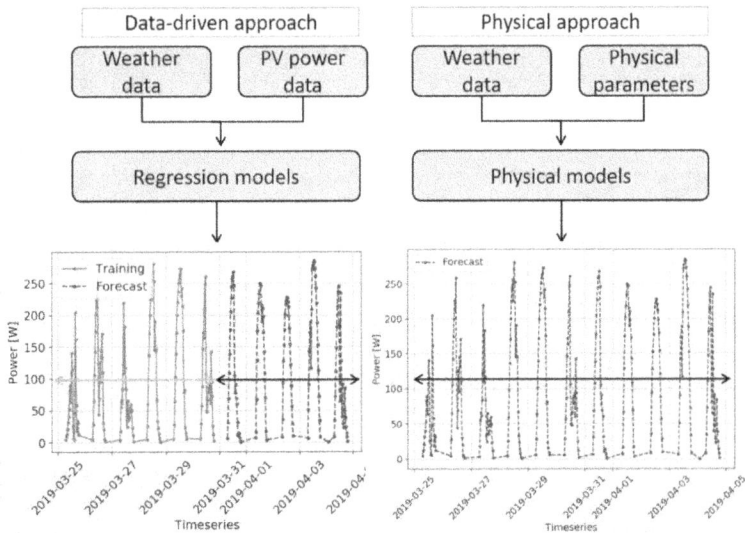

Figure 2.
Schematic of data-driven and physical approaches for PV power prediction. Data-driven approach requires historical data (in green) to train and physical approach requires physical parameters as inputs.

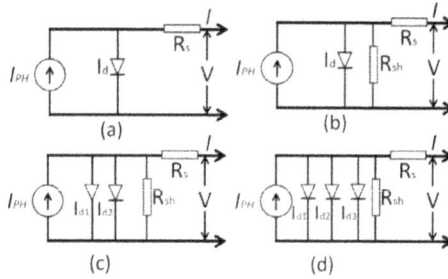

Figure 3.
a, four parameters (4-p) one-diode model. b, five parameters (5-p) one-diode model. c, two-diode model (seven parameters) and d, three diode model (nine parameters).

Model	Parameters	Characteristic
3-p One-diode model	I_{PH} [a], I_{01} [b], n_1 [c]	Basic model No series and shunt resistive losses
4-p One-diode model	I_{PH}, I_{01}, n_1, R_s [d]	Includes series losses No shunt losses
5-p One-diode model	$I_{PH}, I_{01}, n_1, R_s,$ $R_s h$ [e]	Includes both series and shunt shunt resistive losses
Two-diode model	$I_{PH}, I_{01}, I_{02}, n_1$ $, n_2, R_s, R_s h$	Two diodes to represent junction recombination Relevant at low irradiance operation of a PV cell
Three-diode model	$I_{PH}, I_{01}, I_{02}, I_{03},$ $n_1, n_2, n_3, R_s, R_s h$	Takes into account grain boundaries and leakage current

[a] *Photocurrent.*
[b] *Reverse saturation current, the subscripts (1, 2 and 3) represents the diode number respectively.*
[c] *diode ideality factor the subscripts (1, 2 and 3) represents the diode number respectively.*
[d] *Series resistance.*
[e] *Shunt resistance.*

Table 1.
Parameters and characteristics of different diode models.

$$I = I_{PH} - I_0 \left[\exp\left(\frac{V + R_s I}{nV_t}\right) - 1 - \left(\frac{V + R_s I}{R_{sh}}\right) \right] \tag{1}$$

where I, I_o and I_{PH} are the generated solar cell current, reverse saturation current and photo-generated current respectively. V, R_s and R_{sh} are the solar cell voltage, series resistance and shunt resistance respectively. n is the ideality factor of the diode, and V_t is the thermal voltage.

To derive the equivalent circuit model for a PV module, the basic assumption that a PV module comprises of identical PV cells in series is usually taken [6]. This assumption implies that under similar conditions (irradiance and temperature), all the PV cells should generate equal current and voltage. According to [6], the PV module equivalent model is derived from a PV cell diode model (Eq. (1)) as:

$$I_M = I_{PH} - I_0 \left[\exp\left(\frac{V_M + R_s N_s I_M}{nN_s V_t}\right) - 1 - \left(\frac{V_M + R_s N_s I_M}{N_s R_{sh}}\right) \right] \tag{2}$$

where N_s is the total number of cells in series I_M and V_M are the current and voltage of a PV module respectively.

The photo-generated current I_{PH} has a direct relation with solar irradiance and operating solar cell temperature (see Eq. (3) It can be evaluated as [10]:

$$I_{PH} = I_{SC} \frac{G}{G_{STC}} + k_i(T_c - T_{STC}) \quad (3)$$

where G is the given irradiance level, T_c is the cell temperature. G_{STC} and T_{STC} are the irradiance and temperature at standard test conditions (STC) respectively. k_i is the temperature coefficient of the current in (A^0C) and I_{SC} is the short-circuit current at STC.

The reverse saturation current (I_0) can be evaluated as a function of short-circuit current (I_{SC}), open-circuit voltage (V_{oc}), temperature and energy bandgap of a semiconductor (E_g) as:

$$I_o = \left(\frac{T_c}{T_{STC}}\right)^3 \cdot \frac{I_{SC} \exp\left(E_{go}/V_{to} - E_g/V_t\right)}{\exp\left(V_{oc}/nN_sV_{to} - 1\right)} \quad (4)$$

where V_{to} is the thermal voltage at STC and E_{go} is the energy bandgap at T = 0 K.

Readers are referred to [6, 10–12] for extended knowledge on how to derive and evaluate the different PV cell and module model parameters. The functions are also implement is freely available pvlib simulation packages [13].

2.1.2 Temperature dependence of I-V characteristics

Addition to solar irradiation, the I-V curve characteristics also depend on the operating temperatures of the cell T_c. According to IEC60891 standards [14], the temperature and irradiance correction of I-V characteristics are given as:

$$I_{sc} = I_{SC-STC} \cdot \left(\frac{G}{G_{STC}}\right) \cdot (1 + \alpha_{sc}(T_c - T_{STC})) \quad (5)$$

$$V_{OC} = V_{OC-STC} \cdot \left(1 + \beta_{oc}(T_c - T_{STC}) + n.N_s.V_t. \ln\left(\frac{G}{G_{STC}}\right)\right) \quad (6)$$

where α_{sc} and β_{oc} are the temperature coefficients for short-circuit current and open-circuit voltage respectively. n, N_s and V_t are the ideality factor, total number of cells and thermal voltage respectively.

Figure 4 shows the effect of irradiance and module temperature on I_{sc} and V_{oc}. The irradiance has a greater impact on the short-circuit current and the temperature has a greater impact on the open circuit voltage.

According to module mounting (e.g Open rack, close to the roof, insulated rack) and module construction (e.g glass–glass or glass-backsheet), the cell temperature might be some degrees hotter than module temperature [15]. In [15] the cell temperature (T_c) is derived from the module temperature measured at the surface of the module(T_m) and the irradiance G by a simple relations as:

$$T_c = T_m + \frac{G}{G_{STC}} \cdot \Delta T \quad (7)$$

where ΔT is the temperature difference between the cell and the module back surface at an irradiance level of 1000 W/m^2. In [15], ΔT was found to be around 3 ^0C for open-rack mount, 1 ^0C roof mount and 0 ^0C for insulated back.

The module temperature is calculated from solar irradiation, ambient temperatures and/or wind speed using different methods [16]. The commonly used models are: Standard NOCT model (Eq. (8)), the Faiman model (Eq. (9)) [17] and the Kings model also known as Sandia model (Eq. (10)) [15].

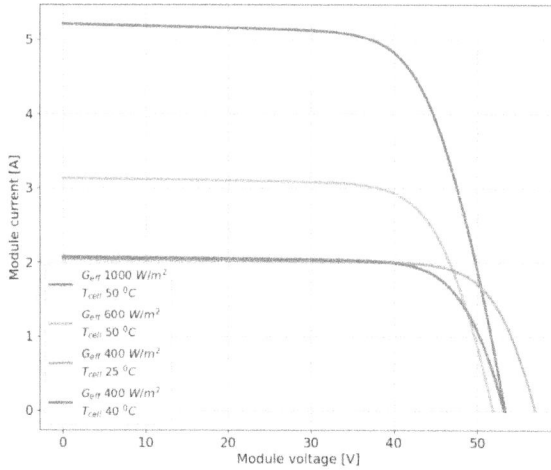

Figure 4.
Effect of irradiance and cell temperature on I-V characteristics. (I-V curve simulated using Pvlib one-diode model).

$$T_m = T_{amb} + \frac{(NOCT - T_{NOCT})}{G_{NOCT}}.G \tag{8}$$

$$T_m = T_{amb} + \frac{G}{U_0 + U_1.WS} \tag{9}$$

$$T_m = T_{amb} + G.\exp(a + b.WS) \tag{10}$$

where NOCT is the nominal operating cell temperature at given conditions (T_{NOCT} = 20 $^\circ C$, G_{NOCT} = 800 W/m^2, wind speed (WS) = 1 m/s), NOCT is in the range of 40–50$^\circ C$ [9]. T_{amb} [$^\circ C$] is ambient temperature, G [W/m^2] is the incident solar irradiance on the module, and WS [m/s] is the wind speed. U_0 $\left[W/^\circ Cm^2\right]$ and U_1 $\left[Ws/^\circ Cm^2\right]$ are the coefficients describing the effect of the radiation on the module temperature and the cooling by the wind, [18] respectively. a and b are parameters that depend on the module construction, materials and on the mounting configuration of the module [15].

2.1.3 Example of PV power prediction using physical method

In this example, we demonstrate a practical application of the described physical model to predict four days PV module power using measured irradiance. The predicted power is then compared with real measured power of the PV module. The properties, electrical and thermal parameters of the PV module are presented in **Table 2**. The module is exposed in Gran Canaria, Spain at tilt angle of 23^0 and azimuth angle of 169^0. The module is installed as open rack-fixed configuration. The power and module temperatures are recorded every 5 minutes. The module temperature is recorded using a Pt100 sensor attached at center-back of the module. In addition, weather data (ambient temperature, wind speed and global horizontal and in plane of array solar irradiation) are also recorded with a 1 minute resolution.

The PV module specific parameters and the weather variables (irradiance and cell temperature) are used as input variable in Pvlib to simulate the five parameter (5-P) one-diode model (Eq. (2)). **Figure 5** shows the simulated I-V curves at

Module properties	
Design and cell technology	Glass–Glass and Poly-crystalline silicon cells
Number of cells	80
Electrical parameters	**Values[a]**
Maximum power rating (P_{mpp})	283 [Wp]
Rated current (I_{mpp})	7.2 [A]
Rated voltage (V_{mpp})	39.3 [V]
Short-circuit current (I_{SC})	7.8 [A]
Open-circuit voltage (V_{OC})	48.9 [V]
Thhermal parameters	**Quantity**
Temperature coefficient of power (γ)	−0.47 [%/K]
Temperature coefficient of short-circuit current (α)	2.39 [mA/K]
Temperature coefficient of open-circuit voltage (β)	−161 [mV/K]
Normal operating cell temperature T_{NOCT}	48 [^{0}C]

[a]*The values are measured at STC (i.e. 1000 W/m^2 irradiance, 1.5 air mass and at 25°C temperature).*

Table 2.
Module properties and manufacturer datasheet electrical and thermal parameters.

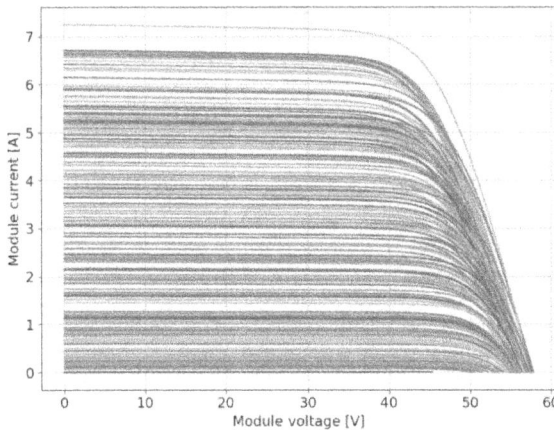

Figure 5.
I-V curves of simulated solar module at different incident irradiance and temperature levels. Irradiance range from 0 w/m^2 to 1168 w/m^2 and module temperature range from 10 °C to 50.3 °C.

different irradiance and temperature levels for a period of four days. 15 minutes aggregated data of temperature and irradiance are used hence the curves are evaluated every 15 minutes. From each I-V curve the power at maximum power point can be computed using:

$$P_{mpp} = I_{mpp} \times V_{mpp} \tag{11}$$

In this example, the uncertainty of power prediction due to temperature models are evaluated. The three commonly used temperature models presented in (Eqs. (8), (9) & (10)) are used to model the module temperature. The parameters of the models are: a = −3.87 & b = −0.0594 for Kings model and $U_0 = 25.6$ &

$U_1 = 25.6$ for Faiman model. These parameters are extracted from literature values in [15, 18] with small modifications to minimize the uncertainty.

To evaluate the uncertainty in prediction, the normalized root mean square error (NRMSE) (Eq. (12)) is used. Additionally, the normalized mean bias error (NMBE) (Eq. (13)) is also evaluated as a metric to capture the average bias in the prediction (i.e, to check whether the predictions are overestimated or underestimated).

$$NRMSE = 100 \times \frac{\sqrt{\frac{1}{N}\sum_{j=1}^{N}\left(p_j - m_j\right)^2}}{\overline{m}} \tag{12}$$

$$NMBE = 100 \times \frac{\frac{1}{N}\sum_{j=1}^{N}\left(p_j - m_j\right)}{\overline{m}}, \quad \overline{m} \neq 0 \tag{13}$$

Where p is the predicted data, m measured data, $\overline{m}$ is the mean of the measured data.

Figure 6 shows the plot of measured and modeled temperature (a), measured and predicted power (b). The uncertainty in temperature models as well as the corresponding uncertainty in power predictions are presented in (c). It is clearly visible that, for each temperature model, the prediction is different and hence the uncertainty value. Generally, the Kings model showed the best performance both in temperature modeling as well as power prediction depending on the NRMSE and R^2 values. The standard NOCT model showed the least performance which is not surprising since the model doesn't take into account the impact of wind speed. All

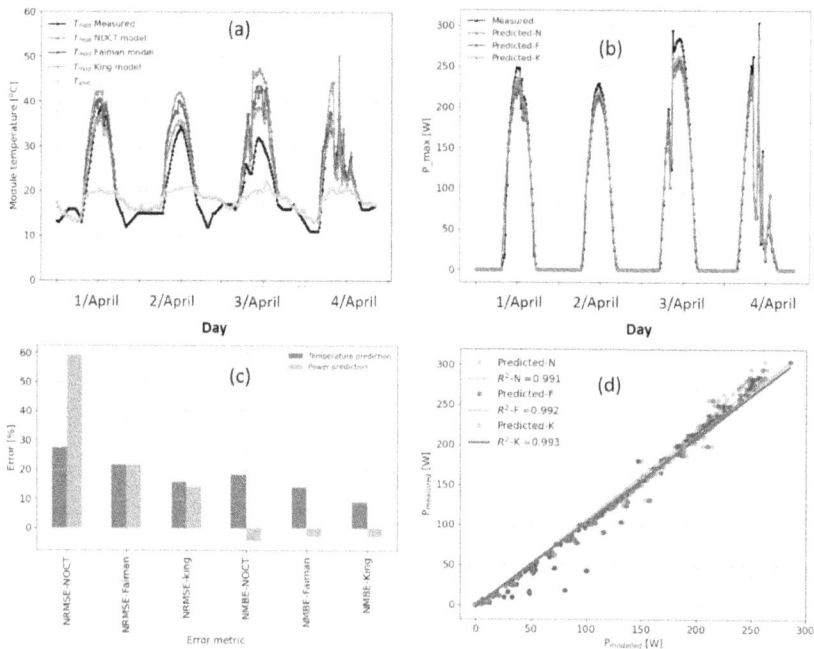

Figure 6.
a, Measured module (black) and ambient (yellow) temperatures, modeled temperature with standard NOCT (green), Faiman (blue) and Kings (red) models. b, Measured (black) and predicted power with module temperature calculated using NOCT (green), Faiman (blue) and Kings (red) models. c, Evaluated NRMSE and NMBE for module temperature modeling (blue) and power prediction (orange). d, Measured versus predicted power with module temperature calculated using NOCT, Faiman and Kings models.

the models overestimate the module temperature which correlates with the underestimation of the predicted power (see **Figure 6c**). It should be noted that, although the Kings model showed good predictions based on this example, it is not enough to guarantee that this will always be the case when the model is applied on different dataset. This is because the accuracy of the temperature models has been found to be influenced by; geographical locations, model design and mounting conditions [18].

2.2 Data-driven methods

Data-driven methods can be categorized into: data-driven heuristic methods, statistical methods and machine learning methods.

2.2.1 Data-driven heuristic methods

The physical models described in Section 2.1 have one big drawback that they require too many input variables which are not usually directly available. In this case, heuristic models are proposed to reduce the number of required inputs. They are heuristic models because they are not developed from physical assumptions/theories. Therefore, they have no physical dependencies/interpretations. They are classified as data-driven models because they are derived from correlation between weather and power output data. In [19], several heuristic models are presented and compared. They are developed on similar principles of generating power forecast from irradiance and module temperature but only differs in the numbers of fitting model parameters. The basic advantage of heuristic models is their simplicity to derive the model parameters from PV power historic data. Here we present the two- (Eq. (14)) and three- (Eq. (15)) parameter models described in [19, 20] respectively.

$$P_{mpp} = \left[\left(1 + x.\ln\left(\frac{G}{G_{STC}}\right) + y.\ln^2\left(\frac{G}{G_{STC}}\right) \right).P_{STC}.\frac{G}{G_{STC}} \right].(1 + \gamma(T_c - T_{STC})$$

(14)

$$P_{mpp} = \left[a.G + b.G + c.G^2.\ln^2\left(\frac{G}{G_{STC}}\right) \right].(1 + \gamma(T_c - T_{STC})$$

(15)

Where P_{mpp}, is the generated power at maximum power point by a PV module, G is the simulated or measured irradiance T_c is the cell temperature evaluated using equation (Eq. (7)), a, b, c, x and y are the models fitting parameters γ is the temperature coefficient of power in (%/^{0}C). P_{STC} and $G_{STC} = 1000w/m^2$ are the power and the irradiance at STC.

2.2.2 Example of PV power prediction by heuristic methods

In this example, the described heuristic models are applied to predict the power of the same PV module described in subSection **2.1.3**. To calibrate the models, six days historical data (25/March −31/March) are used. The extracted parameters are presented in **Table 3**. To demonstrate the usefulness of temperature correction term, the second term in (Eq. (15)) is removed and the model is re-calibrated.

In **Table 3** 3-parameter model corresponds to calibration without a temperature term and 3-parameter-T_{corr} model corresponds to calibration with a temperature correction term.

Model	Parameter				
	a	b	c	x	y
1. parameter model (Eq. (14))	—	—	—	0.0255	−0.03016
2. parameter model (Eq. (15))	0.2842	−2.935e-5	−0.0106	—	—
3. parameter-T_{corr}[a] model (Eq. (15))	0.2802	−8.985e-7	−0.0111	—	—

[a]T_{corr} is the temperature correction.

Table 3.
Extracted model parameters of the heuristic models.

Figure 7a shows a plot of the four days measured power and the predict power using 2-parameters and the 3-parameters model. **Figure 7b** shows the uncertainty in model calibration and the corresponding uncertainty in prediction using the different models. According to the NRMSE and NMBE values, the 3-parameter model with temperature correction term showed the best performance. From the same figure it can also be concluded that it is important to include a temperature correction in power prediction since the same model showed the least performance when applied without a temperature correction.

2.2.3 Statistical and Machine learning methods

Like heuristic models, statistical and machine learning (ML) methods are also based on historical data to generate PV power forecast. While heuristic models focus on an in-deep formulation of mathematical operations, statistical models require selecting a model that considers previous knowledge of the system. ML methods require the selection of a predictive algorithm by relying on its empirical capabilities. Statistical models aim to "inference" the outcome of a model, while ML approaches aim to find generalizable predictive patterns [21]. Both statistical and ML methods are data-driven approaches that rely on the availability and accuracy of existing operational data to generate the forecasting. Usually, the larger the historical data, the better the PV system can be understood in terms of operational behaviour under different weather conditions and hence the better the forecasting accuracy.

The list of published methods is extensive and a case-to-case benchmark is usually needed. Statistical methods such as Naive method, ARIMA (Autoregressive

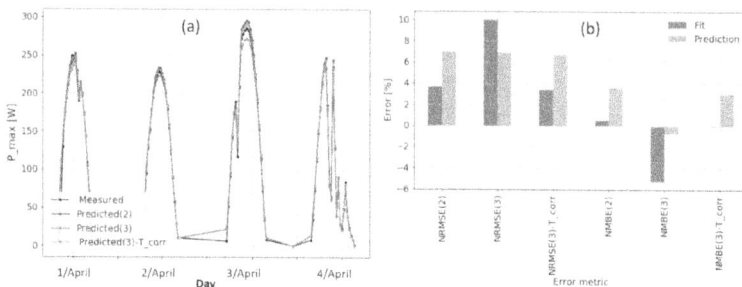

Figure 7.
a, Measured power (black), predicted power using 2-parameter model (blue) and with 3-parameter model without (green) and with (red) temperature correction. b, Evaluated NRMSE and NMBE for the models during calibration (blue) and prediction (orange).

Integrated Moving Average), SARIMA (seasonal ARIMA), Ordinary Least Squares (OLS) or Facebook Prophet (FbP) are typically applied for PV forecasting with and without regressors [22–24]. Below is a basic description of the commonly applied statistical methods. A detailed description of each method can be found on the respective cited reference.

- **Autoregressive Integrated Moving Average (ARIMA)**: This method is composed of three main elements: the auto-regression order (p), differencing order (d) and moving average order (q). The statistical formulation of this method can be found in [25].

- **Seasonal ARIMA (SARIMA)**: This model is an extension of the ARIMA approach, which adds the seasonal behaviour of the dataset analysed. This feature is of interest for PV applications due to the high seasonality on daily and annual basis observed in PV systems [26].

- **Ordinary Least Squares (OLS)**: This method analyses the system by fitting linear relationships between one or more input variables, and by minimizing the sum of square errors of a continuous or at least interval outcome variable (actual versus predicted values) [27].

- **Facebook Prophet (FbP)**: This methodology has been developed to allows non-experts in data science to adapt and configure the model to their needs. The FbP method is based on a decomposable time series model including trend, seasonality, holidays (not important for our application), and an error term. It is also possible to define the type of evolution: linear or logistic. For PV forecasting, it facilitates the modelling for short- and long-term by enabling features such as time resolutions and temporal seasonalities [28].

More advanced methods, the so-called machine learning (ML) methods, can provide better results [29, 30], however, in most cases they require more computational efforts. Some examples of machine learning methods used in PV power forecasting include [2, 29–31]: k-nearest neighbors (k-NN), artificial neural networks (ANN), support vector machine (SVM), random forests (RF) and light gradient boost machines (LightGBM). The basic description of these methods is presented below with the references for a detailed description.

- **k-Nearest Neighbors (k-NN)**: Is classified as one of the simplest and straightforward ML method. The K-NN algorithm compares the current state's Euclidean distances with training samples in feature space to select the "k" nearest neighbours used in predictions. Detailed description and application in PV power forecasting are presented in [32, 33]

- **Artificial Neural Networks (ANN)**: Inspired by biological neural networks, this algorithm is composed by neurons (mathematical units) and weights (the link between mathematical units). Using gradient-based optimization techniques, the ANNs learn a specific task (e.g., prediction) by the optimization of the "weights". This method is widely used in PV power forecasting [2] described in [29, 34, 35]

- **Support Vector Machine (SVM)**: The method separates the data linearly and transforms it into a higher dimensional feature space through a specific kernel function. The linear separation is performed with the so-called "hyperplanes".

The SVMs can be used for regression as well as classification tasks. In [32] the model is presented and applied for short-term PV power forecasting.

- **Random Forests (RF)**: The RF algorithm is based on an ensemble learning. A set of decision/regression trees is created and the final result is voted. For a regression problem, a set of regression trees are trained and the forecast will equal to the mean of individual regression tree results [36].

- **Light gradient boosting machine (LightGBM)**: This algorithm is an advanced gradient boosting decision tree (GBDT), which combines two techniques to find more effectively (higher accuracy and high processing speed) the optimal split point in the GBDT: (1) Gradient-based one-side sampling (GOSS), to reduce the number of data instances, and (2) exclusive feature bundling (EFB) to reduce the feature space. In [30, 37], the method is presented and applied for PV power prediction.

ML algorithms can be classified as supervised and unsupervised. A supervised ML algorithm uses labelled training data. It is related to a standard fitting procedure to find the unknown function/relationship between the input and output variables. The unsupervised ML algorithm uses unlabelled training data to find the data patterns (e.g., in the samples' clustering). For PV power prediction, the supervised algorithms are commonly used, due to weather forecasting availability. In general, the procedure to run a ML algorithm can be composed of the following stages:

- Data collection: the available historical data (weather and PV operational) are collected and filtered. The collection of weather forecasting data is also considered.

- Feature selection: identification of the most relevant variables with regard to the PV power output selected for further analysis.

- Data augmentation: in this stage, the enhancement of the initial dataset is expected by typically applying mathematical operations (e.g., physical relationships) to one or more relevant input variables.

- Dataset split: the input dataset is divided into a training and testing dataset. Also, a validation dataset is recommended. This task is typically applied over the sorted or random timestamps.

- Accuracy improvement: statistical indicators such as the MBE, RMSE or R^2 are used to quantify the accuracy of any forecasting model. Cross-validation techniques are recommended.

A simple ANN network architecture is shown in **Figure 8**, where the layers of a multilayer perceptron (MLP) for PV power forecasting are presented. The input data for training can be the historical weather and PV power output, while for testing and forecasting, expected weather variables are the input to the expected PV power output in the future.

2.2.3.1 Example of PV power prediction by statistical and machine learning approaches

The statistical and ML approaches are applied to the same dataset used for physical and heuristic models. To train the models, the regressors selected are

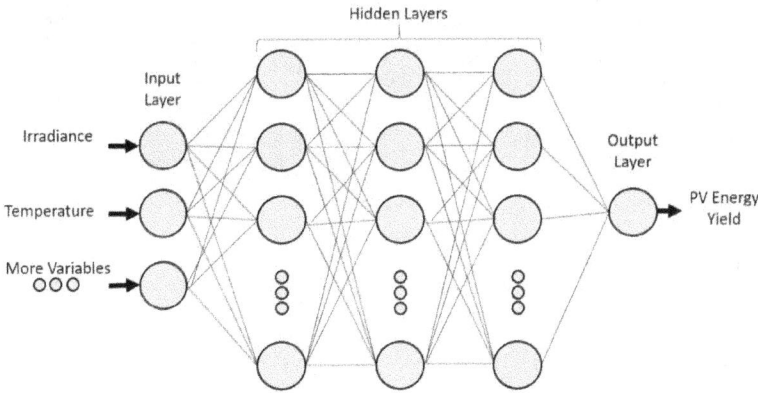

Figure 8.
Multilayer perceptron (MLP) for PV power forecasting, where the input layer includes at least irradiance and temperature, while the output layer comprises the PV energy yield.

Model		Parameter	
	Setup 1	**Setup 2**	**Setup 3**
ARIMA[a]	p: 0	q:0	d: 0
SARIMA[a]	p: 1	q:1	d: 1
OLS[a]	—	—	—
Prophet[b]	Daily Seasonality	Changepoint prior scale: 0.01	—
DNN[c]	Hidden layer size: (40,20,10)	relu activation	adam solver
LightGBM[d]	Number of leaves: 10	Min. data in leaf: 5	Number iterations: 100
SVM[c]	Kernel: rbf	Degree: 3	Regularization: 1e6

[a] *Statsmodels.*
[b] *Prophet.*
[c] *Sklearn.*
[d] *Lightgbm.*

Table 4.
Setup parameters for different statistical and machine learning approaches.

limited to the plane-of-array irradiance and PV module temperature, while the target variable is the PV output power. Most of these models are already implemented in open-source software packages (i.e., statsmodels [38], prophet [28], sklearn [39] and lightgbm [39]) and executed with Python scripts. Setting parameters for each statistical and machine learning method are given in **Table 4**.

Figure 9a shows a plot of the four-day measured power and the predicted power using statistical and machine learning models. **Figure 9b** shows the uncertainty in model calibration and the corresponding uncertainty in prediction using the different models. According to the NRMSE, LightGBM model shows the best performance, followed by the SVM and Facebook Prophet.

Generally, comparing the uncertainty values among the physical, heuristic, statistical and machine learning methods (see **Figures 6c, 7b** and **9b**), physical method has the worst performance in comparison to other methods. This is not surprising given that physical models generate prediction without preliminary performance data, unlike the counterpart models that base their predictions from historical data. Also, the different assumptions and the too many input parameters increase the uncertainty range in physical models.

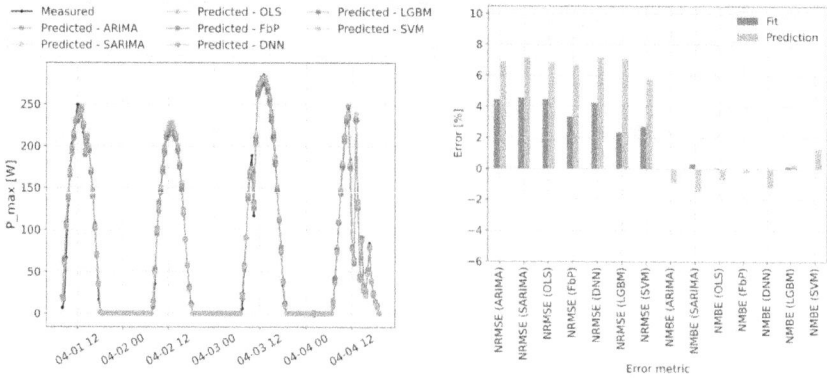

Figure 9.
A, Measured power (black), predicted power using statistical and machine learning approaches. B, Evaluated NRMSE and NMBE of statistical and machine learning approaches at calibration and predictions.

3. Lifetime PV power forecasting

To begin with, it is important to understand how the lifetime of a PV module is defined. Unlike other electrotechnical devices where the term lifetime is clearly defined [40], the definition of a PV module "lifetime" is somewhat complex. This is because, despite the catastrophic events (such as fire) it is unlikely that a PV module drops its power generation to zero. However, even though a PV module is still generating power, its output might be too low to be economically viable to continue its operation. Therefore, in general terms PV lifetime is defined in economical rather than technical terms.

For economical viability of PV projects, most PV module manufacturers guarantee a power reduction of less than 20% within 25–30 years of operation. The 20% power reduction is usually referenced at standard test conditions (STC) (modules tested under 25^0C temperatures, 1000 W/m^2,irradiance and air mass 1.5). Therefore, in this context the lifetime of a PV modules is defined as the time required for a PV module to loss it's STC power by 20% .

The actual performance of a PV module throughout its lifetime is very uncertain and difficult to accurately forecast. This is because many factors can influence the performance of a PV module. Some of these factors may include: solar resource, the quality of the PV components and the long-term variations in system performance (degradation). All these factors increases the uncertainty in PV lifetime forecast. The Internation Energy Agency- Photovoltaic Power Systems Programme (IEA-PVPS) -task 13 report [41] provides a detailed overview of the uncertainties in lifetime yield predictions. To improve the accuracy and to achieve reliable lifetime PV forecast, all these effects must be explored separately. Here, we asses the effect of degradation on lifetime PV power forecasting.

3.1 Effect of degradation on lifetime PV power forecasting

Degradation is defined as the gradual and non-reversible decrease in PV performance over time. Degradation is a crucial influencing factor to be taken into account during lifetime PV power forecast. This is because over time the PV components are ageing and deteriorating in their normal operation. Understanding how PV degrades is a very widely studied topic in the PV community but it is also among the not well understood topics. This can be explained by the numerous factors influencing PV degradation. These factors include: PV technology, bill of materials (BoM), climatic conditions [4], transportation, installation and operational conditions. More-so, since new materials are being proposed frequently, it increases the complexity to correlate the factors that influence PV degradation. Usually, different materials have different degradation kinetics and are influenced by different stress factors differently.

In lifetime power prediction models, degradation effect is included in a number of ways. For example, according to the survey carried out among PVPS-task 13 experts regarding degradation effects in lifetime energy yield prediction, the following assumptions are taken [41]: (a). A variable degradation during the first five year of operation and a fixed degradation from 5 to 30 years of operation. A degradation of 1–2% is assumed in the first year, 0.7% to 0.5% to year 5 and0.3% to 0.5% up to year 30. (b). Initial degradation of 0.3% to 1.0% in the first years to include the effects of initial degradation modes such as light induced degradation (LID). (c). Constant degradation over the years with the exception of the first year to take into account technology specific behaviour.

Generally, a constant degradation rate with linear performance loss is considered (see Eq. (16)). However, some authors [42–44], have evaluated and modeled the non-linearity in degradation rates and performance. For example in [43] a non-linear power degradation model (Eq. (17)) was proposed and applied in [44] with a time-dependent degradation rates to predict PV performance lifetime.

$$P(t) = P_{mpp} \cdot (1 - k.t) \tag{16}$$

$$P(t) = P_{mpp} \times \left[1 - exp\left(-\left(\frac{B}{k.t}\right)^{\mu} \right) \right] \tag{17}$$

Where P_{mpp} is the power calculated using either the physical model or the heuristic models described in the previous sections. k [%/year] is the degradation rate B and μ are model and shape parameters respectively.

3.2 Example of lifetime PV power forecasting

In this example, the effect of degradation rate on lifetime power forecast is presented. Using, 30 years of historical weather data (global irradiance, ambient temperature and wind speed from ERA 5 reanalysis [45]), three different degradation scenarios are presented and their impact on lifetime power and energy yield prediction. The first scenario is using a non-linear performance degradation with a shape parameter ($\mu = 1.2$), the second scenario is using a linear performance degradation and the third scenario is using a non-linear model with a different shape parameter ($\mu = 0.2$). It should be noted that in all the three cases, a constant degradation rate of 0.8%/year corresponding to a lifetime of 25 years (to have a − 20% power loss) is used. In all the cases P_{mpp} is calculated using a 3-parameter heuristic model (Eq. (15)). Although it is not usually the case, the prediction without degradation effect and its impact on lifetime energy prediction is also shown in this example.

Figure 10 shows the PV lifetime power and yield predictions using different scenarios. It can be seen that depending on the degradation scenario, the lifetime yield is significantly different. In numbers, when compared to the usually used linear scenario, a relative difference of over 5.0% is evaluated in respect with the non-linear scenarios.

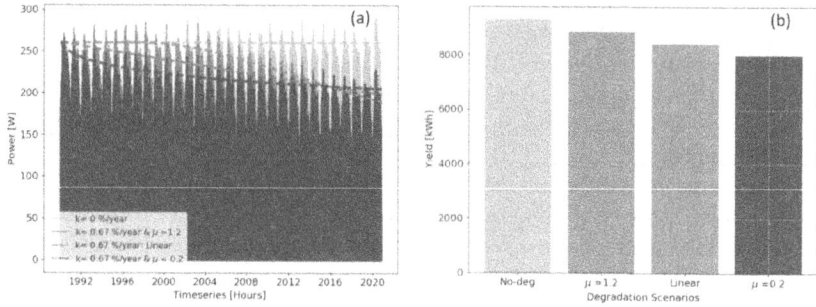

Figure 10.
A, Lifetime power prediction using different degradation scenarios: non-linear with shape parameter = 1.2 (red), linear (green) and non-linear with shape parameter = 0.2 (blue) as well as a no degradation scenario (cyan). B, Corresponding lifetime yield for all the respective scenarios.

4. Conclusions

PV power forecasting is important to stabilize the electrical grids, financing PV projects and also to plan operational and maintenance activities. In this regard, different methods are proposed to forecast the PV power generation. In this chapter, the different methods used in PV power forecasting are presented, applied and their accuracy in PV forecasting is evaluated using measured PV module and weather data. The degradation effect on lifetime PV power forecasting is also assessed using two main scenarios; linear degradation scenario and non-linear degradation scenario. The key observations in the chapter are:

- The uncertainties in PV module temperature modelling affect the forecasting accuracy. In the chapter, three temperature models: Standard NOCT, Faiman and King's are compared, the King model showed the best performance among the three models. The standard NOCT model, that neglects the impact of wind speed displayed huge uncertainty.

- Data-driven models outperforms physical models in prediction accuracy. This can be explained by the fact that physical models are derived from to many assumptions and that they need too many input parameters that are usually approximated.

- For lifetime PV power forecasting, a relative difference of over 5% is evaluated between the linear and non-linear degradation scenarios.

Conflict of interest

The authors declare no conflict of interest.

Author details

Ismail Kaaya[1*] and Julián Ascencio-Vásquez[2,3]

1 Fraunhofer ISE, Freiburg, Germany

2 Faculty of Electrical Engineering, University of Ljubljana, Ljubljana, Slovenia

3 3E sa, Brussels, Belgium

*Address all correspondence to: ismail.kaaya@ise.fraunhofer.de

IntechOpen

References

[1] Global solar photovoltaic capacity [Internet]. Available from: https://www.globaldata.com/global-solar-photovoltaic-capacity-expected-to-exceed-1500gw-by-2030-says-globaldata/ [Accessed: 28-October-2020].

[2] Antonanzasa J, Osoriob N, Escobar R, Urraca R, Martinez-de-Pisona F.J, Antonanzas-Torresa F: Review of photovoltaic power forecasting. Solar energy. 2016; 136(15): 78–111. https://www.sciencedirect.com/science/article/abs/pii/S0038092X1630250X [Accessed: 10 November 2020]

[3] Pelland, Sophie, Remund, Jan, Kleissl, Jan, Oozeki, Takashi and De Brabandere, Karel Photovoltaic and Solar Forecasting: State of the Art. (IEA-PVPS T14-01: 2013) , International Energy Agency Photovoltaic Power Systems Programme (2013). [Online]: https://iea-pvps.org/wp-content/uploads/2013/10/Photovoltaic_and_Solar_Forecasting_State_of_the_Art_REPORT_PVPS__T14_01_2013.pdf [Accessed: 5-January-2021].

[4] Ascencio-Vasquez, J., Kaaya, I., Brecl, K., Weiss, K.-A., & Topic, M., Global Climate Data Processing and Mapping of Degradation Mechanisms and Degradation Rates of PV Modules. Energies, 2019, 12, 4749. doi:10.3390/en12244749.

[5] Nazmul Islam Sarkar Md: Effect of various model parameters on solar photovoltaic cell simulation: a SPICE analysis. Renewables. 2016; 3–13. DOI: 10.1186/s40807-016-0035-3

[6] Tian, H., et al.: A cell-to-module-to-array detailed model for photovoltaic panels. Solar Energy, 2012. 86(9): 2695–2706. DOI:10.1016/j.solener.2012.06.004

[7] PVsyst [Internet]. Available from: https://www.pvsyst.com/features/ [Accessed: 16-November -2020]

[8] PVWATTs [Internet]. Available from: https://pvwatts.nrel.gov/ [Accessed: 16-November -2020]

[9] Alberto Dolara, Sonia Leva, Giampaolo Manzolini:Comparison of different physical models for PV power output prediction. Solar energy. 2015; 119(-): 83–99. DOI:10.1016/j.solener.2015.06.017

[10] One-Diode Model [Internet]. Available from: https://www.sciencedirect.com/topics/engineering/one-diode-model [Accessed: 20-November -2020]

[11] De Soto W, Klein S.A, Beckman W. A: "Improvement and validation of a model for photovoltaic array performance", Solar Energy, 2006; 80 (1): 78–88. DOI:10.1016/j.solener.2005.06.01

[12] Single Diode Equivalent Circuit Models [Internet]. https://pvpmc.sandia.gov/modeling-steps/2-dc-module-iv/diode-equivalent-circuit-models/ [Accessed: 20-November -2020]

[13] Pvlib python [Internet]. https://pvlib-python.readthedocs.io/en/stable/index.html [Accessed: 20-November -2020]

[14] International Electrotechnical Vocabulary. Chapter 191: Dependability and Quality of Service, IEC60050-191, International Electrotechnical Comission, Geneva, CH, Standard, 1990.

[15] Kratochvil, Jay A, Boyson, William Earl, and King, David L. Photovoltaic array performance model.. United States: N. p., 2004. Web. doi:10.2172/919131

[16] Segado P.M , Carretero J & Sidrach-de-Cardona, M: Models to predict the operating temperature of different

photovoltaic modules in outdoor conditions. Progress in Photovoltaics: Research and Applications, 2015, V23, 1267–1282. DOI:10.1002/pip.2549.

[17] Faiman, D. Assessing the outdoor operating temperature of photovoltaic modules. Progress in Photovoltaics: Research and Applications, 2008; 16(4), 307–315. https://doi.org/10.1002/pip.813

[18] Koehl, M., Heck, M., & Wiesmeier, S: Categorization of weathering stresses for photovoltaic modules. Energy Science & Engineering, 2018,V6, 93–111. DOI:10.1002/ese3.622189.

[19] Ding K, Ye Z, Reindl T. Comparison of Parameterisation Models for the Estimation of the Maximum Power Output of PV Modules. Energy Procedia, 2012; V25, 101–107. DOI: 10.1016/j.egypro.2012.07.014

[20] Performance Evaluation and Prediction of BIPV Systems under Partial Shading Conditions Using Normalized Efficiency; Energies 2019; V12, 3777; doi:10.3390/en12193777

[21] Bzdok D, Altman N & Krzywinski N, Statistics versus machine learning, Nat Methods, 2018 15, 233–234. DOI: urlhttps://doi.org/10.1038/nmeth.4642

[22] Subhra Das, Short term forecasting of solar radiation and power output of 89.6kWp solar PV power plant, Materials Today: Proceedings, 2020, DOI: 10.1016/j.matpr.2020.08.449.

[23] David P. Larson, Lukas Nonnenmacher, Carlos F.M. Coimbra, Day-ahead forecasting of solar power output from photovoltaic plants in the American Southwest, Renewable Energy, 2016 V.91, 11–20, DOI: 10.1016/j.renene.2016.01.039.

[24] M. Bouzerdoum, A. Mellit, A. Massi Pavan, A hybrid model

(SARIMA–SVM) for short-term power forecasting of a small-scale grid-connected photovoltaic plant, Solar Energy, 2013, 98, 226-235, DOI: 10.1016/j.solener.2013.10.002.

[25] Pasari S, Shah A, Time Series Auto-Regressive Integrated Moving Average Model for Renewable Energy Forecasting. In: Sangwan K., Herrmann C. (eds) Enhancing Future Skills and Entrepreneurship. Sustainable Production, Life Cycle Engineering and Management: Springer; 2020. p.71–77. DOI: https://doi.org/10.1007/978-3-030-44248-4_7

[26] Dimri T, Ahmad S & Sharif M, Time series analysis of climate variables using seasonal ARIMA approach. J Earth Syst Sci,2020, 129, 149. DOI: https://doi.org/10.1007/s12040-020-01408-x

[27] Zdaniuk B, Ordinary Least-Squares (OLS) Model. In: Michalos A.C. (eds) Encyclopedia of Quality of Life and Well-Being Research. Springer, Dordrecht, 2014, DOI: https://doi.org/10.1007/978-94-007-0753-5_2008

[28] Taylor SJ, Letham B. Forecasting at scale. PeerJ Preprints 5:e3190v2, 2017; [Online] https://doi.org/10.7287/peerj.preprints.3190v2

[29] Spyros Theocharides, George Makrides, Andreas Livera, Marios Theristis, Paris Kaimakis, George E. Georghiou, Day-ahead photovoltaic power production forecasting methodology based on machine learning and statistical post-processing, Applied Energy, 2020; V 268, 115023, DOI: 10.1016/j.apenergy.2020.115023.

[30] Julian Ascencio-Vasquez, Jakob Bevc, Kristjan Reba, Kristijan Brecl, Marko Jankovec and Marko Topic, Advanced PV Performance Modelling Based on Different Levels of Irradiance Data Accuracy, Energies 2020, 13(9), 2166; DIO, 10.3390/en13092166

[31] Suresh V, Janik P, Rezmer J, Leonowicz Z. Forecasting Solar PV Output Using Convolutional Neural Networks with a Sliding Window Algorithm. Energies. 2020; 13(3):723.

[32] Fei Wang, Zhao Zhen, Bo Wang, and Zengqiang, Comparative Study on KNN and SVM Based Weather Classification Models for Day Ahead Short Term Solar PV Power Forecasting, applied science, 2018, 8, 28; DOI: 10.3390/app8010028

[33] Zhang, Y.; Wang, J. GEFCom2014 Probabilistic Solar Power Forecasting based on k-Nearest Neighbor and Kernel Density Estimator. In Proceedings of the 2015 IEEE Power & Energy Society General Meeting; 26–30 July 2015; Denver,CO, USA; DOI: 10.1109/PESGM.2015.7285696

[34] Utpal Kumar Das, Kok Soon Tey, Mehdi Seyedmahmoudian, Saad Mekhilef, Moh Yamani Idna Idris, Willem Van Deventer, Bend Horan, Alex Stojcevski, Forecasting of photovoltaic power generation and model optimization: A review, Renewable and Sustainable Energy Reviews, 2018; 81(1), p.912–928 DOI: https://doi.org/10.1016/j.rser.2017.08.017.

[35] Guido Cervone, Laura Clemente-Harding, Stefano Alessandrini, Luca Delle Monache, Short-term photovoltaic power forecasting using Artificial Neural Networks and an Analog Ensemble,Renewable Energy,2017, v108, p.274–286, DOI:https://doi.org/10.1016/j.renene.2017.02.052.

[36] Huertas Tato, J.; Centeno Brito, M. Using Smart Persistence and Random Forests to Predict Photovoltaic Energy Production. Energies 2019, 12, 100. DOI: https://doi.org/10.3390/en12010100

[37] Caroline Persson, Peder Bacher, Takahiro Shiga, Henrik Madsen,

Multi-site solar power forecasting using gradient boosted regression trees, Solar Energy, 2017; v.150, p.423–436, DOI: https://doi.org/10.1016/j.solener.2017.04.066.

[38] Seabold, Skipper, and Josef Perktold. "Statsmodels: Econometric and statistical modeling with python." Proceedings of the 9th Python in Science Conference. 2010.

[39] Scikit-learn: Machine Learning in Python, Pedregosa et al., Journal of Machine Learning Research 12 2011; 12, 2825–2830.[Online] http://scikit-learn.sourceforge.net.

[40] IEC60891, 2010. Photovoltaic Devices. Procedures for Temperature and Irradiance Corrections to Measured IV Characteristics. IEC60891. [Internet] https://standards.globalspec.com/std/1207301/IEC%2060891 [Accessed: 23-November-2020].

[41] Reise C, Müller B, Moser D, and et al, IEA PVPS Task 13: Uncertainties in PV System Yield Predictions and Assessments (2018). [Online]: https://iea-pvps.org/key-topics/uncertainties-in-pv-system-yield-predictions-and-assessments/ [Accessed: 16-November-2020].

[42] Marios Theristis, Andreas Livera, C. Birk Jones, George Makrides, George E. Georghiou, and Joshua S. Stein, Nonlinear Photovoltaic Degradation Rates: Modelingand Comparison Against Conventional Methods, IEEE Journal of Photovoltaics, 2020; 10(4):1112–1118, DOI: 10.1109/JPHOTOV.2020.2992432

[43] Kaaya I., Koehl M.,Mehilli A., Mariano S. d. C., & Weiss K. A. 'Modeling Outdoor Service Lifetime Prediction of PV Modules: Effects of Combined Climatic583Stressors on PV Module Power Degradation', IEEE Journal of Photovoltaics, 2019; 9, 1105–1112. doi:10.1109/JPHOTOV.2019.291619

[44] Kaaya, I., Lindig, S., Weiss, K.-A., Virtuani, A., Ortin, M. S. d. C., & Moser, D., Photovoltaic lifetime forecast model based on degradation patterns. Progress in Photovoltaics: Research and Applications, 2020; 28, 979–992. doi: 10.1002/pip.3280

[45] Copernicus Climate Change Service (C3S) (2017): ERA5: Fifth generation of ECMWF atmospheric reanalyses of the global climate. Copernicus Climate Change Service Climate Data Store (CDS), [Internet] Available from: https://cds.climate.copernicus.eu/cdsapp#!/home, [Accessed: 16-November -2020]

Chapter 10

Concentrator Photovoltaic System (CPV): Maximum Power Point Techniques (MPPT) Design and Performance

Olfa Bel Hadj Brahim Kechiche and Habib Sammouda

Abstract

The research carried out in this work aimed to study the performance of MPPT techniques applied to the Concentrator Photovoltaic (CPV) System for the research and the pursuit of the Maximum Power Point (MPP).This study presents a modeling and simulation of the CPV system. It consists of a PV module located in the focal area of a parabolic concentrator, a DC / DC converter (Boost), two MPPT controls (P&O and FL) and a resistive load. This chapter presents the two MPPT techniques (P&O and FL) performances. The obtained results show the importance of cooling systems integration with CPV system. This hybrid system design results in good MPPT P&O and FL performance. The numerical results obtained with Matlab/Simulink® software have generally shown that the two MPPT controls result in better performance in terms of speed, and accuracy, stability. In fact they showed that the CPV system is stable.

Keywords: Concentrator photovoltaic System (CPV), Converter DC-DC (Boost), MPPT Techniques, Performances, Perturb & Observe (P&O) algorithm, Fuzzy Logic (FL) algorithm, Matlab/Simulink®

1. Introduction

Today, Concentrator Photovoltaic (CPV) systems are among the important technologies for converting solar radiation into electrical energy. Despite the high cost of this technique, the CPV system attracted attention last years many researcher for their high power output compared with conventional module systems. Santosh Kumar Sharma et al. [1] designed the aspects and the performance of a rooftop grid-connected solar photovoltaic power plant (RTGCSPVPP). The RTGCSPVPP is installed at Gauri Maternity Home Ramkrishna Puram Kota Rajasthan, India for supplying the energy to whole hospital building. T. Mrabti, et al. [2] presented the implementation and operation of the first installation prototype high concentration photovoltaic (CPV) in Morocco. This installation is formed by three two-axis sun trackers connected to the national electricity grid. In fact, they showed the first experimental results concerning the electrical operation of this plant and its daily energy production as a function of meteorological conditions.

On the other hand, photovoltaic modules are expensive and their electrical characteristics suffer from climatic variations, it is therefore necessary to extract the maximum power to increase the efficiency of the module [3]. A.Saxena et al. [4] evaluated the non-linear I-V characteristics of a photovoltaic solar module and its maximum power point which depends on climatic conditions (temperature and irradiation).

Additional, the PV module efficiency is limited for two reasons: first, part of the solar radiation is converted into heat. Second, the module temperature increases during the energy production. Therefore, the use of a cooling system becomes necessary. Sanjeev et al. [5] presented the various cooling technologies available for CPV systems and they showed that cooling systems can provide an uniform and low cell temperature.

Also, there are many techniques called MPPT (Maximum Power Point Tracking) [6]. The most common MPPT methods are Perturb & Observe (P&O) and the Incrementation of Conductance (INC). Other MPPT algorithms include the use of a Fuzzy Logic Controller (FLC), an Artificial Neural Network (ANN), [7–9].

D. Djalel, et al. [10] showed the MPPT techniques (P&O and Fuzzy logic) performance under STC or Standard Test Conditions, which correspond to irradiation G of 1 kW/m^2 at spectral distribution of AM1.5 and a cell temperature T of 25°C. Then they carried out a comparison between these two MPPT controls. According to the simulation results, the fuzzy logic method generates good performance: low oscillating, more stable operating point than P&O and important precision to operate the MPP. M. A. Enany et al. [10] have modeled and simulated same MPPT techniques such as: ANFIS, FCO, Fuzzy logic, Increment of conductance, Disturbance and P&O observation. Then they compared between these techniques. And they concluded that the ANFIS method and fuzzy logic control present the best performance.

The previous studies mentioned below do not take into consideration the photovoltaic concentration conditions. To our knowledge, the MPPT techniques performance in these conditions has rarely been studied in the open literature. In order to further the study of CPV systems, improvements have been made to the present study, including the integration of the cooling system with adequate temperature and the evaluation of the performance behavior of the commercial PV module.

The purpose of this chapter is to compare the performances of two MPPT techniques P&O and FL for a CPV system in the aim to determine the suitable technique.

This chapter is organized as follows. Part 2 describes the modeling a PV module placed at the focus of a parabolic concentrator. Part 3 presents the improvement of a proposed CPV module with a cooling system, then the simulation of this global system consisting of a CPV module, a boost converter, two MPPT algorithms (P&O and FL) and a resistive DC / DC load. Part 4 presents numerical results and a comparison between the two MPPT techniques. Finally, Part 4 concludes this work.

2. Modeling a PV module placed at the focus of a parabolic concentrator

In order to achieve a higher efficiency of a PV module, we propose to place it in the focal space of a concentrator composed by a double reflective parabolic concentrator, **Figure 1**.

This system is composed by:

- *A first reflector:* is a heliostat as a sun tracking system with a reflection coefficient equal to 1.

- **A second reflector:** is a parable that is composed of a set of curved mirrors. Its role is to reflect and focus the light received by the heliostat on a receiver placed in the focal space of the parabolic concentrator.

- **A receiver:** is a fixed photovoltaic module that concentrates the received radiation.

Figure 1 shows the block diagram of the proposed photovoltaic system. This system is composed by the following elements:

- A PV module placed in the focal space of a concentrator

- A DC/DC converter Boost type

- Resistive load

- And an MPPT controller

In the state of solar concentration, the output current module, denoted I_{PV}, is given by (Eq. (1)), [11]:

$$I_{PV} = N_p I_{ph} - N_p I_s \left[\exp \left(\frac{\frac{V_{PV}}{N_s} + \frac{IR_s}{N_p}}{nV_{th}} \right) - 1 \right] - \frac{\left(\frac{N_p V_{PV}}{N_s} + IR_s \right)}{R_{sh}} \quad (1)$$

The photo current I_{ph} is mainly depending on the incident irradiance and the cell operating temperature. It can determine using (Eqs. (2) and (3)), [12]:

$$I_{ph} = \frac{G}{G_{ref}} \left(I_{sc,ref} + K_i.\Delta T \right) \quad (2)$$

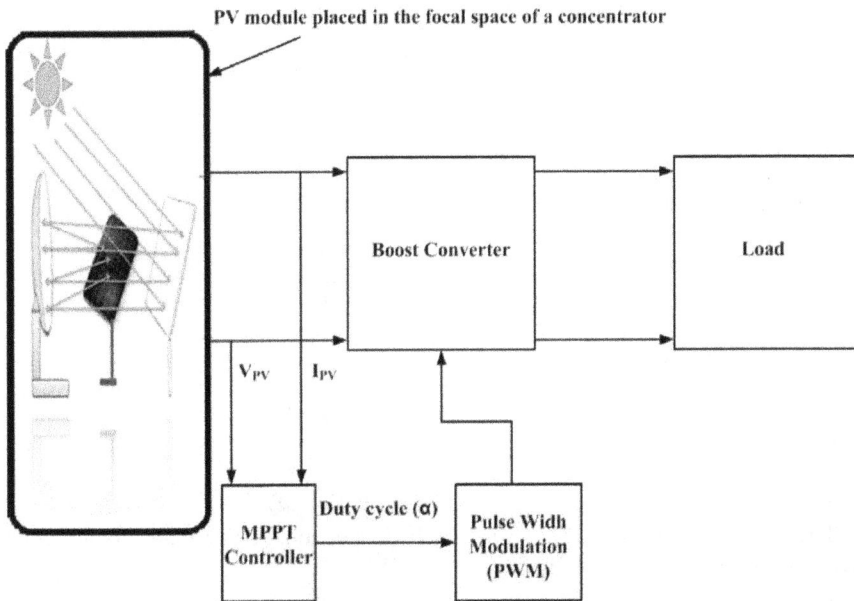

Figure 1.
Proposed CPV system.

where

$$\Delta T = T - T_{ref} \tag{3}$$

The cell operating temperature T varies with the incident irradiance, which is described by (Eq. (4)), [13]:

$$T = T_{amb} + \left(\frac{T_{NOCT} - 20}{800}\right) G \tag{4}$$

The diode saturation current I_s at any operating conditions is related to its reference conditions by the following equation, [7]:

$$I_s = I_{s,ref} \left(\frac{T}{T_{ref}}\right)^3 \exp\left[\frac{qE_g}{nK}\left(\frac{1}{T_{ref}} - \frac{1}{T}\right)\right] \tag{5}$$

The reverse saturation current at STC condition $I_{s,ref}$ is depending on open circuit voltage (V_{oc}) and can be calculated by (Eq. (6)), [12]:

$$I_{s,ref} = \frac{I_{sc}}{\exp\left(\frac{V_{oc}}{nV_{th}}\right) - 1} \tag{6}$$

The material band gap energy E_g is obtained by (Eq. (7)) using Varshni relation, [6, 14].

$$E_g(T) = E_g(0) + \frac{\alpha T^2}{T + \beta} \tag{7}$$

Table 1 E_{g0}, α and β silicon parameters [13]:

	E_{g0} (T = 0 K)	$\alpha.10^{-4}$, eV/K^2	β, K
Si	1.17	4.73	636

Table 1.
The E_{g0}, α and β silicon parameters

Then, the Si band gap as a function operating temperature is determined by (Eq. (8))

$$E_g(T) = 1.17 + \left(\frac{4.73 \cdot 10^{-4}T^2}{T + 636}\right) \tag{8}$$

The series resistor module R_s can be approximately expressed by (Eq. (9)), [15]:

$$R_s = R_{s,ref} - \left[\frac{n}{I_s} \exp\left(\frac{-V_{oc}}{n}\right)\right] \tag{9}$$

$R_{s,ref}$ is the module series resistor measured at STC (Ω)

The shunt resistor module R_{sh} is inversely proportional to irradiance incident on the CPV module and is given by (Eq. (10)), [15]:

$$R_{sh} = R_{sh,ref}\left(\frac{G_{ref}}{G}\right) = R_{sh,ref}\left(\frac{1}{C}\right) \tag{10}$$

where the concentration ratio C is defined by (Eq. (11)):

$$C = \frac{G}{G_{ref}} \tag{11}$$

$R_{s,ref}$ is the module shunt resistor measured at STC (Ω)

The diode ideality factor n is considered according to $C = \frac{G}{G_{ref}}$ as function of cell operating temperature and reference cell temperature, [15]:

$$n = n_{ref} \frac{T}{T_{ref}} \tag{12}$$

For Si-poly, n_{ref} = 1.3 is the diode ideality factor at STC, [13]

The thermal voltage of the cell V_{th} is defined by (Eq. (13)):

$$V_{th} = \frac{KT}{q} \tag{13}$$

K is the Boltzmann constant, 1.38×10^{-23}J/K, q is the Electron charge, 1.602×10^{-19}C.

3. CPV system configuration improvement

To improve the CPV system performance, the PV module temperature must be reduced. Hence the interest of inserting a heat sinks. Thus we will assemble the concentrator with a cooling system below the PV module to maintain the value of its temperature constant.

An active dissipation exchanger will be used to maintain the module temperature at 35°C. **Figure 2** represents the modification made to the PV module, [16, 17].

SOLKAR make 36- Watt, Photovoltaic module is taken as the reference module for simulation and the manufacturer specifications details are given in **Table 2**.

The module series resistor and the module shunt resistor of SOLKAR Photovoltaic Module are supposed ideal by, [2] and are fixed successively at $R_{s,ref} = 0.001\Omega$ and $R_{sh,ref} = \infty$.

Based on (Eq. (1)), the solar module model was implemented in MATLAB/Simulink® environment.

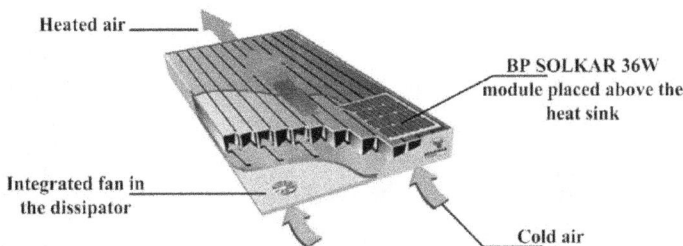

Figure 2.
Heat sink placed below the PV module under the solar concentration condition.

Maximum Power P_m	37.08W
Voltage at Maximum power V_m	16.56V
Current at Maximum power I_m	2.25A
Open circuit voltage V_{oc}	21.24V
Short circuit current I_{sc}	2.55A
Number of series Cells N_s	36

Table 2.
SOLKAR datasheet values at STC.

4. Boost converter model

Figure 3 shows the boost converter structure used in this chapter. The boost converter is composed with a MOSFET and Diode switching elements where are supposed to be ideal, a resistor, inductance and capacitor where are supposed to be linear, time invariant and frequency independent, [13].
The average output voltage V_c is given by:

$$V_c = \frac{V_{PV}}{1 - \alpha} \tag{14}$$

where
$L = 290\mu H$, $C1 = 250\mu F$, $C2 = 330\mu F$, $R = 35\Omega$ and the PWM frequency $f_{PWM} = 10kHz$.

5. MPPT scheme

The MPPT algorithm used the measured values of the output voltage and/or the output current of the PV module to estimate the duty cycle (D) of the DC–DC converter in order to keep the electrical load characteristics with those of the PV module at the Maximum Power Point MPP, [13].

5.1 Perturb & observe (P&O) algorithm

P&O algorithm is most popular and usually adopted strategy between all MPPT techniques. This algorithm is frequently used for commercial PV module because it is easy to implement and inexpensive, [9, 17].

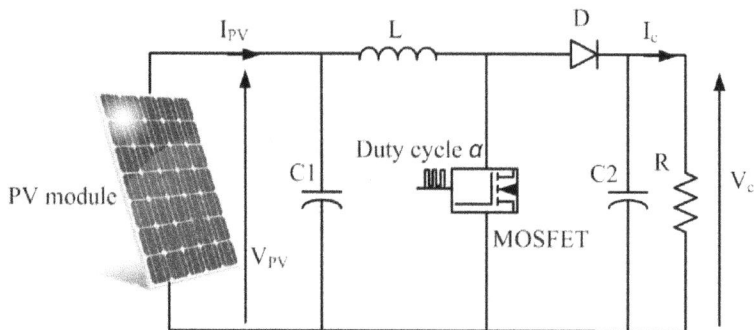

Figure 3.
Boost converter structure.

The P&O method is based on, [15–17]:

- Periodical measuring the PV module voltage $V(k)$ and current $I(k)$ to calculate the output power $P(k)$;

- Perturbing (increasing or decreasing) the switching duty cycle (D) of the Boost converter to change the operating point. In this study a slight perturbation $(\Delta D = 0.01)$ is introduced in the system.

- Observing the output power variation $\Delta P = P(k) - P(k - 1)$:

  - If $\Delta P > 0$, the Maximum Power Point MPP will be approached, therefore the perturbation should be kept the same for the following stage;

  - Otherwise the perturbation should be reversed.

- This process is repeated until the MPP is reached.

Figure 4 presents the P&O algorithm implemented in Matlab/Simulink®.

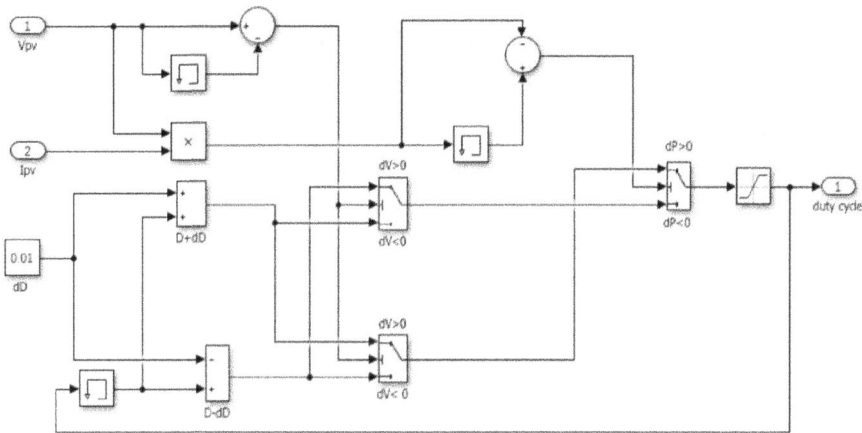

Figure 4.
P&O algorithm in MATLAB/Simulink®.

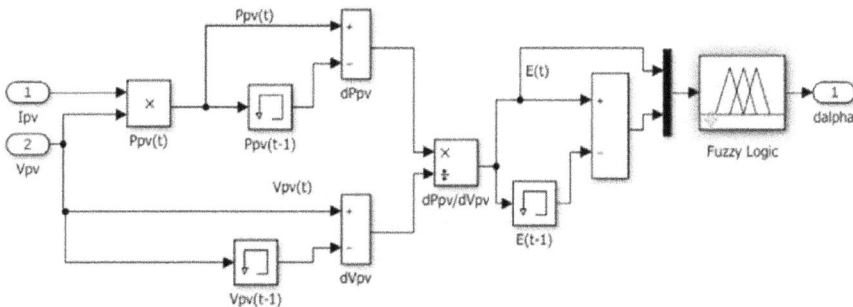

Figure 5.
Fuzzy logic algorithm in MATLAB/Simulink®.

5.2 Fuzzy logic (FL) algorithm

The FL algorithm checks the output power value of the PV module at each instant (t) and then calculates the power variation $\left(\frac{dP}{dt}\right)$ according to the voltage variation, [16, 18].

The fuzzy logic algorithm generally consists of three stages: the fuzzification, the rules and the defuzzification, [16, 18].

Figure 5 illustrate the fuzzy logic (FL) algorithm implanted in Simulink environment.

6. Results and discussion

6.1 MPPT control performance under the concentration conditions

In the first part of this subsection, the concentration ratio is fixed to C = 1x. For this report, the PV module temperature simulated by the software Matlab/Simulink® is equal to T = 53.75 °C.

The simulation results of the CPV system using two different techniques (P&O and FL) are presented successively by the **Figures 6–8**:

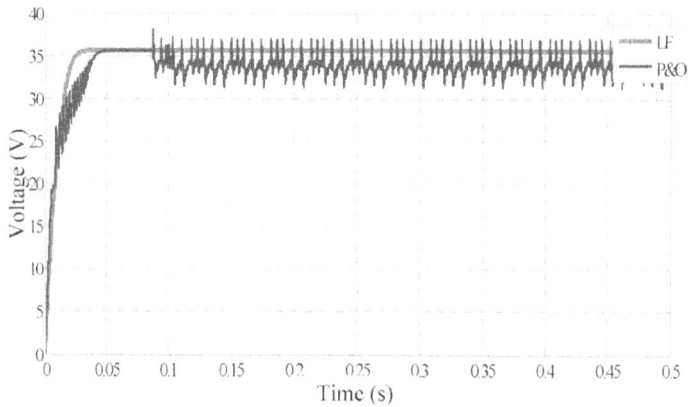

Figure 6.
Output voltage using the MPPT control (P&O and FL) for C = 1x and T = 53.75°C.

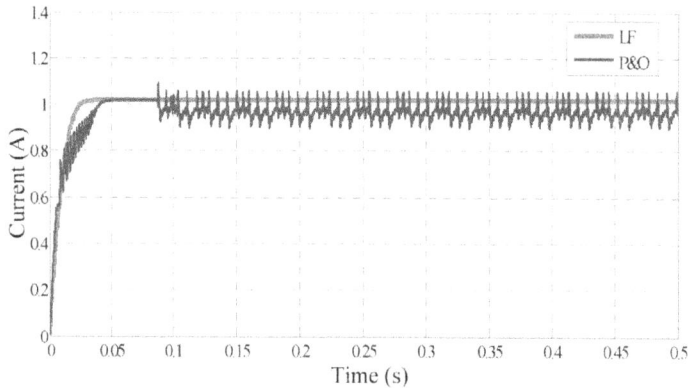

Figure 7.
Output current using the MPPT control (P&O and FL) for C = 1x and T = 53.75°C.

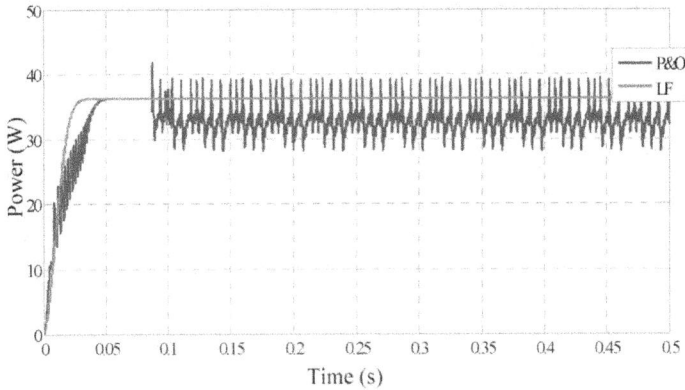

Figure 8.
Output power using the MPPT control (P &O and FL) for C = 1x and T = 53.75°C.

Concentration report MPPT		C = 1x	C = 2x	C = 3x
Simulated temperature		*T = 53.75°C*	*T = 87.5°C*	*T = 121°C*
Fuzzy logic (LF)	I_s (A)	0.97	1.088	1.126
	V_s (V)	33.72	38.08	39.39
	P_s (W)	32.48	41.42	44.34
	η_{mppt}(%)	96	96	96
Perturbation and observation (P&O)	I_s (A)	0.963	1.088	1.126
	V_s (V)	33.71	38.08	39.39
	P_s (W)	32.46	41.42	44.34
	η_{mpp} (%)	85.8	96	96

Table 3.
V_s, I_s, P_s and η_{mppt} variation of the MPPT control (P&O and FL) as a function of the concentration ratio.

Then, the CPV system performance parameters, the output voltage V_s, the output current I_s, the maximum output power P_s and the efficiency η_{mppt} for different values of the solar concentration ratio (1x, 2x, 3 x) are determined in **Table 3**.

From the results obtained, it can be seen that the "Fuzzy Logic" control does not exhibit oscillations at the steady state of the current curve I_s, the voltage V_s and the power P_s and that the response time of this technique is fast. While, the P&O control exhibits several disturbances due to climate change (temperature and concentration) and results in a longer response time than the other technique. For a concentration ratio C = 1x, the efficiency of the CPV system using the FL control is equal to 75% while the efficiency of the CPV system using the P&O control is equal to 74.1%. For a C > 1x concentration ratio, the efficiency of the CPV system using both FL and P & O controls is stabilized by up to 60%.

So, we can deduce that the FL control performs better than the P&O control.

The characteristics (I-V) and (P-V) of the CPV system using the P&O and LF control are represented successively in **Figures 9** and **10** for different values of the concentration ratio solar (1x, 2x, 3 x).

(a) Characteristics (I-V)

(b) Characteristics (P-V)

(c) Zoom on the PPM

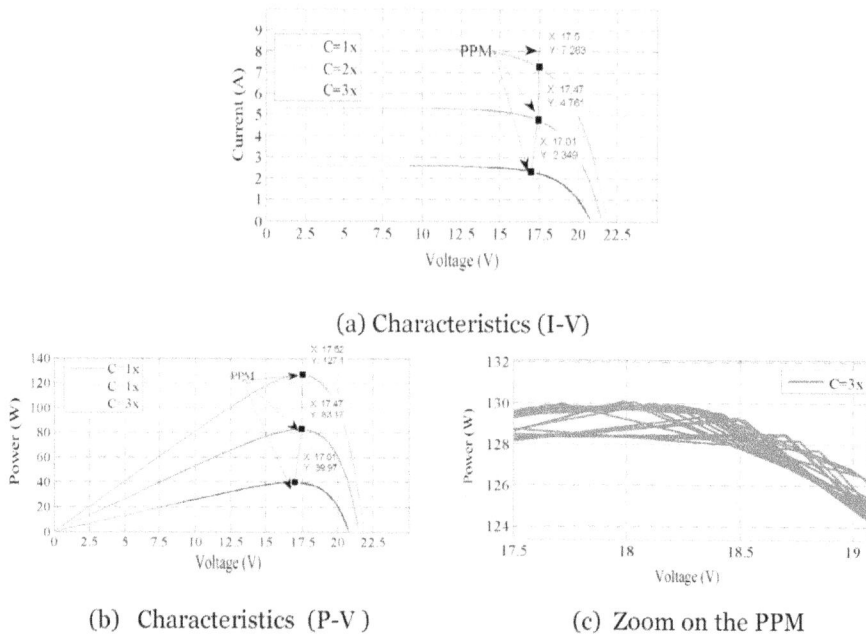

Figure 9.
Characteristics (I-V) and (P-V) of the CPV system using the P&O control under different solar concentration values. (a) Characteristics (I-V). (b) Characteristics (P-V). (c) Zoom on the PPM.

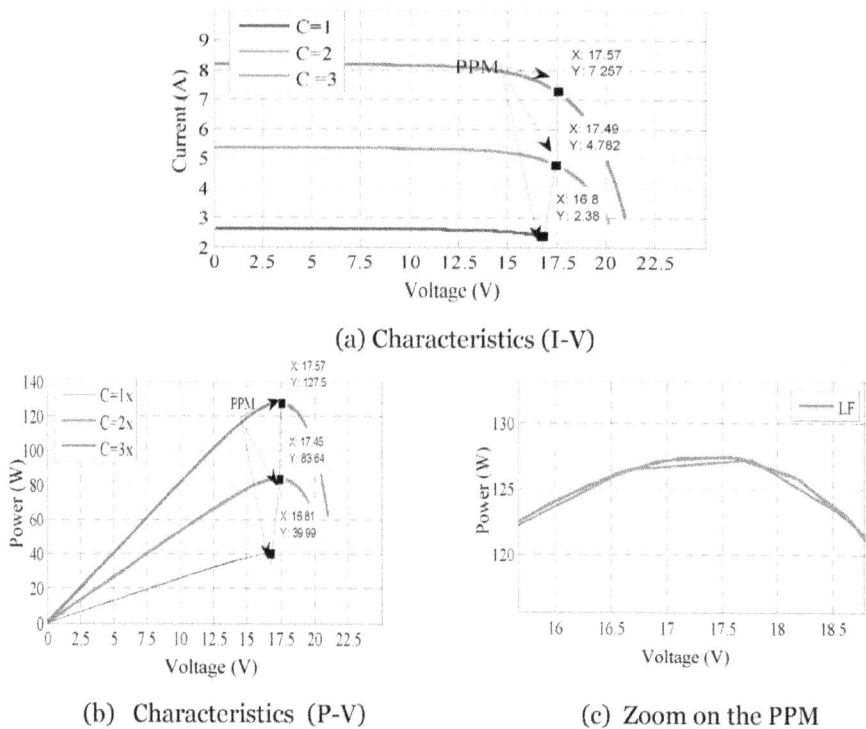

(a) Characteristics (I-V)

(b) Characteristics (P-V)

(c) Zoom on the PPM

Figure 10.
Characteristics (I-V) and (P-V) of the CPV system using the FL control under different solar concentration values. (a) Characteristics (I-V). (b) Characteristics (P-V). (c) Zoom on the PPM.

As shown in **Figures 9** and **10**, it can be seen that the PV module output (I-V) and (P-V) characteristics strongly influenced by the variations in metrological conditions (temperature and concentration) for both control P&O and FL. It should be noted that the maximum power point MPP of the PV module is also influenced by the concentration ratio C and the temperature T.

When the temperature varies, the P&O control shows the existence of strong oscillations around the maximum power point, **Figure 9(c)**. Due to these oscillations around this point, the CPV system shows energy losses.

Contrariwise, during a temperature variation, and using the fuzzy logic control, there are weak oscillations around the MPP which limits the power losses, **Figure 10(c)**.

6.2 MPPT control performance with the improve CPV system

In this section, initially, we maintained the same model under the concentration conditions implemented under Matlab / Simulink® software by setting the temperature at 35°C. Secondly, we varied the solar concentration ratio C, in a range of (2x to 10x), to study the performance of the two MPPT controls used in the CPV system.

6.2.1 P&O control performance

From the output power curves $P_s(t)$, **Figure 11**, it is noted that the increase in concentration causes an increase in power. But also for each power curve, we obtain two parts:

- **Regime1:** it is the transient regime of the power presents enormous peaks. The transient state indicates the control speed.

- **Regime2:** the steady state shows the stability of the power over time.

The output power signal P_s stabilizes in a reduced response time, e.g. for C = 3x, Tr = 0.0106 s. This shows that the FL control performs well its role which is the

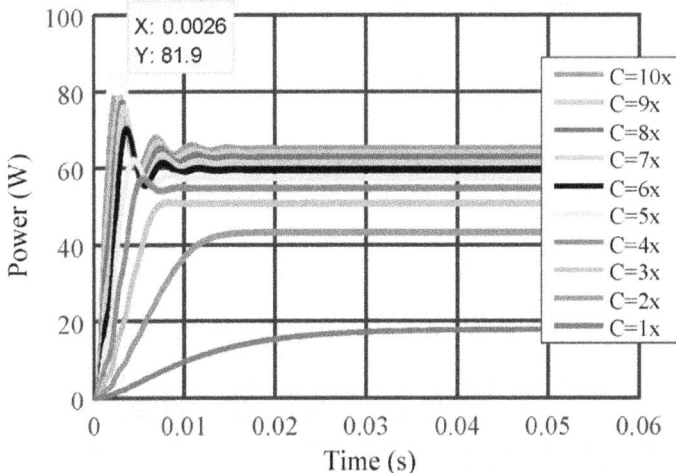

Figure 11.
CPV system output power under the concentration conditions at a constant temperature (35°C) and with P&O control.

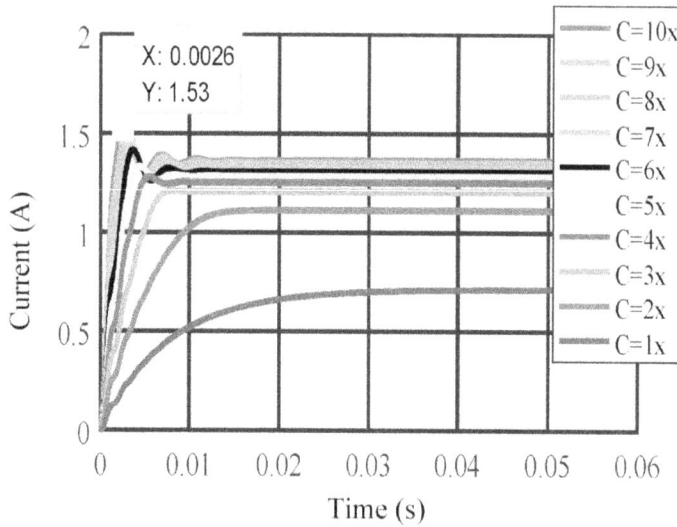

Figure 12.
CPV system output current under the concentration conditions at a constant temperature (35°C) and with P&O control.

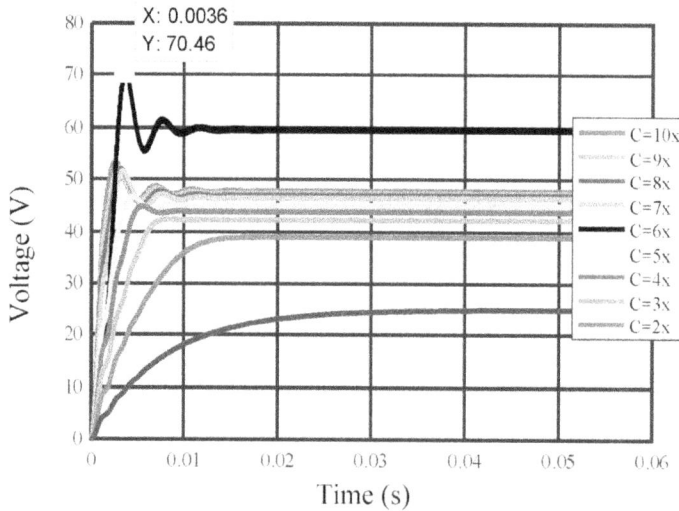

Figure 13.
CPV system output voltage under the concentration conditions at a constant temperature (35°C) and with P&O control.

tracking of the maximum power point on the one hand and secondly, the CPV system output signal is stable.

When C = 10x, the P_s curve has the largest peak (P_s = 65.23 W).

According to **Figures 12** and **13**, the output current I_s and the output voltage V_s have a transient region and a permanent region. Similarly, in the previous results, we note that the transient regime has large peaks.

The Boost converter that ensures the electrical energy transit between the PV module and the resistive load, it is characterized by their impedance which creates voltage drops (disturbances of the duty cycle) and energy losses.

Parameters	C = 1x	C = 2x	C = 3x	C = 4x	C = 5x	C = 6x	C = 7x	C = 8x	C = 9x	C = 10x
Current I$_s$ (A)	0.715	1.114	1.207	1.253	1.284	1.307	1.326	1.341	1.354	1.365
Voltage V$_s$ (V)	25.04	38.98	42.24	43.87	44.95	45.76	46.4	46.93	47.38	47.78
Power P$_s$ (W)	17.92	43.41	50.98	54.99	57.73	59.82	61.51	62.93	64.15	65.23
MPPT efficiency (%)	53	56	59	63	65	67	69	70	72	73
Response time	0.058	0.0173	0.0106	0.0106	0.0121	0.0149	0.015	0.016	0.0196	0.0174

Table 4.
The characteristic quantities of the "SOLKAR 36 W" module under the concentration conditions at a constant temperature (35°C).

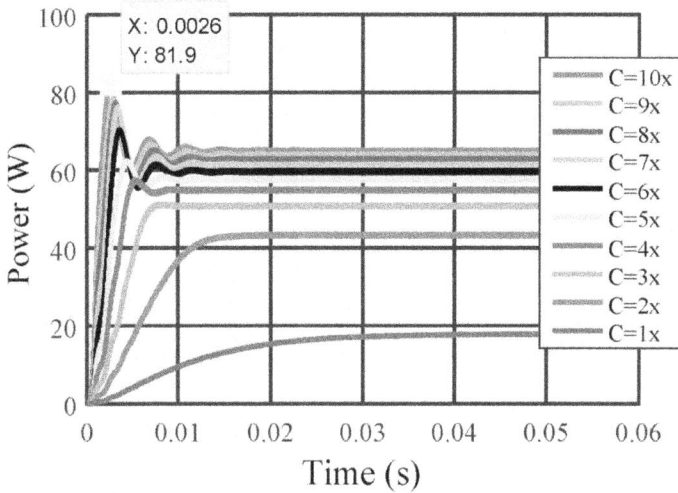

Figure 14.
CPV system output power under the concentration conditions at a constant temperature (35°C) and with FL control.

Strong currents and impedance can cause long-term oscillations.

The simulation results show that this system can adapt to a resistive load (R = 35 Ω). Indeed, it can give a fast response and a good transient performance, insensitive to changes in external disturbances.

Table 4 summarizes the PV module characteristic parameters under the concentration conditions at a constant temperature (35°C): the output voltage Vs, the output current Is, the output power Ps, the MPPT efficiency η$_{mppt}$, and the response time T$_r$.

6.2.2 Fuzzy logic (FL) control performance

From **Figures 14–16**, we note that the results obtained by the FL control are similar to those obtained by the P&O control, the same transient regime which we find the peaks and the same steady state which is stable and the oscillations are gone.

It can be seen that the new configuration of the CPV system has improved the performance of the P&O control. We can therefore deduce that the appearance of oscillations in the old CPV system is due to the rise in temperature. By setting this parameter, it was possible to stabilize the output signals of the system.

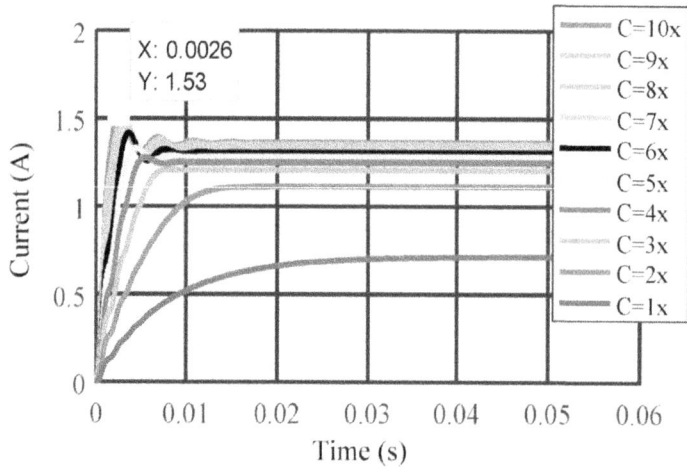

Figure 15.
CPV system output current under the concentration conditions at a constant temperature (35°C) and with FL control.

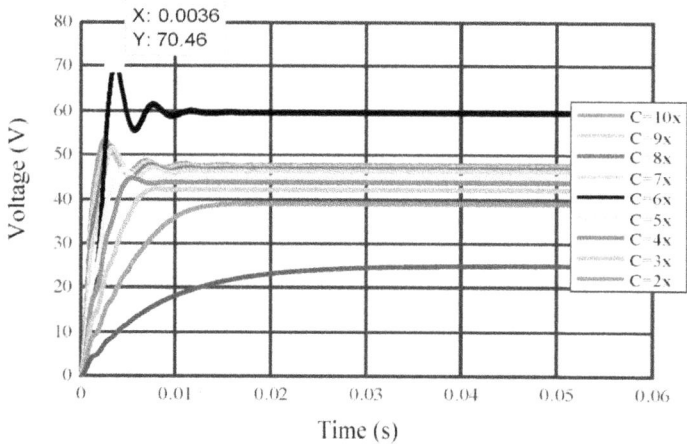

Figure 16.
CPV system output voltage under the concentration conditions at a constant temperature (35°C) and with FL control.

MPPT Type	Stability	Sensors number	Response time (Convergence time)	Digital or analog	% Yield
LF	Stable	1 current 1 voltage	0.0106	digital	73
P&O	Stable	1 current 1 voltage	0.0106	Analog Or Digital Or Both	73

Table 5.
The performances of the two techniques "P&O" and "FL".

The following **Table 5** shows the performance of two MPPT techniques P&O and FL for a CPV system with a cooling system:

From **Table 5**, it can be concluded that the P&O control in the CPV system with a cooling system becomes more interesting than the FL control. Indeed, these two controls have the same evolution of the output signals (P_s, V_s, I_s), same response time, same transient regime and same performance but the advantage of the P&O control and that its practical implementation is simpler than the FL control.

The P&O technique has the following performances:

- Low implantation cost

- the ease of its implementation

- no need for precise inference parameters

In return, the fuzzy logic control in the CPV system has disadvantages such that:

- High implantation cost

- the complexity of its implementation

- Need precise inference parameters.

7. Conclusion

This work aims to present the principle of a CPV system, thus to study the modeling of a PV module placed at the focus of a parabolic concentrator. Then, we simulated this CPV system in a Matlab/Simulink ® environment under different conditions of temperature and concentration ratio. Finally we showed the performance of the two MPPT commands (P&O and FL).

Simulation results showed that both MPPT methods (P&O and FL) were successful in continuing and reaching the PPM peak power point although disturbances due to temperature and concentration changes. As well as the control by fuzzy logic causes the best performance in terms of response time, stability and accuracy.

In the second part of this chapter, we improved the CPV system configuration by adding a cooling system and setting the temperature to 35°C. The simulations results in these new conditions show that the performances of the two MPPT P&O and FL controls are identical and the oscillations are thus due to the rise in temperature.

Acknowledgements

This project was supported by the Tunisian Ministry of Higher Education and Scientific Research under Grant LabEM – ESSTHSousse – LR11ES34.

Conflict of interest

The authors declare no conflict of interest.

Author details

Olfa Bel Hadj Brahim Kechiche* and Habib Sammouda
Laboratory of Energy and Materials (LR11ES34), High School of Sciences and
Technology of Hammam Sousse, Sousse University, Hammam Sousse, Tunisia

*Address all correspondence to: olfa.belhadjbrahimkechiche@essths.rnu.tn;
belhajbrahimolfa@yahoo.fr

IntechOpen

References

[1] Santosh Kumar Sharma, D. K. Palwalia, and V. Shrivastava, "Performance Analysis of Grid-Connected 10.6 kW (Commercial) Solar PV Power Generation System," *Appl. Sol. Energy (English Transl. Geliotekhnika)*, vol. 55, no. 5, pp. 269–281, 2019, doi: 10.3103/S0003701X19050128.

[2] T. Mrabti *et al.*, "Implantation et fonctionnement de la première installation photovoltaïque à haute concentration 'CPV'au Maroc," *Rev. des Energies Renouvelables*, vol. 15, no. 2, pp. 351–356, 2012.

[3] S. O. F. Dhyia Aidroos Baharoona, Hasimah Abdul Rahmana, Wan Zaidi Wan Omara, "Historical Development of Concentrating Solar Power Technologies to Generate Clean Electricity Efficiently - A review," *Renew. Sustain. Energy Rev. Manuscr.*, vol. 41, pp. 996–1027, 2015.

[4] A. R. Saxena *et al.*, "Performance Analysis of P&O and Incremental Conductance MPPT Algorithms Under Rapidly Changing Weather Conditions," *J. Electr. Syst.*, pp. 292–304, 2014.

[5] S. Jakhar, M. S. Soni, and N. Gakkhar, "Historical and recent development of concentrating photovoltaic cooling technologies," *Renew. Sustain. Energy Rev.*, vol. 60, pp. 41–59, 2016, doi: 10.1016/j.rser.2016.01.083.

[6] Y. P. Huang and S. Y. Hsu, "A performance evaluation model of a high concentration photovoltaic module with a fractional open circuit voltage-based maximum power point tracking algorithm," *Comput. Electr. Eng.*, vol. 51, pp. 331–342, 2016, doi: 10.1016/j.compeleceng.2016.01.009.

[7] S. S. Haq, B. W. S. Sunder, and G. M. Zameer, "Design and simulation of MPPT algorithm of photovoltaic system using intelligent controller," *Int. J. Adv. Sci. Tech. Res.*, vol. 6, no. 3, pp. 337–346, 2013.

[8] O. Singh and S. K. Gupta, "A review on recent Mppt techniques for photovoltaic system," *2018 IEEMA Eng. Infin. Conf. eTechNxT 2018*, pp. 1–6, 2018, doi: 10.1109/ETECHNXT.2018.8385315.

[9] N. Hussein Selman, "Comparison Between Perturb & Observe, Incremental Conductance and Fuzzy Logic MPPT Techniques at Different Weather Conditions," *Int. J. Innov. Res. Sci. Eng. Technol.*, vol. 5, no. 7, pp. 12556–12569, 2016, doi: 10.15680/ijirset.2016.0507069.

[10] D. Dib, M. Mordjaoui, and G. Sihem, "Contribution to the performance of GPV systems by an efficient MPPT control," *Proc. 2015 IEEE Int. Renew. Sustain. Energy Conf. IRSEC 2015*, no. December, 2016, doi: 10.1109/IRSEC.2015.7454930.

[11] A. Zahedi, "Review of modelling details in relation to low-concentration solar concentrating photovoltaic," *Renew. Sustain. Energy Rev.*, vol. 15, no. 3, pp. 1609–1614, 2011, doi: 10.1016/j.rser.2010.11.051.

[12] O. Meriem and A. Haddi, "Comparative study of the MPPT control algorithms for photovoltaic panel," *Proc. Int. Conf. Ind. Eng. Oper. Manag.*, no. September, pp. 1840–1852, 2017.

[13] H. Bellia, "A detailed modeling of photovoltaic module using MATLAB," *NRIAG J. Astron. Geophys.*, 2014, doi: 10.1016/j.nrjag.2014.04.001.

[14] V. Kumar Garg, "a Review Paper on Various Types of Mppt Techniques for Pv System," *Int. J. Eng. Sci. Res.*, vol. 4, no. 5, pp. 320–330, 2014, [Online]. Available: www.ijesr.org.

[15] O. Bel Hadj Brahim Kechiche, B. Barkaoui, M. Hamza, and H. Sammouda, "Simulation and comparison of P&O and fuzzy logic MPPT techniques at different irradiation conditions," *Int. Conf. Green Energy Convers. Syst. GECS 2017*, pp. 1–7, 2017, doi: 10.1109/GECS.2017.8066266.

[16] G. Mittelman, A. Kribus, and A. Dayan, "Solar cooling with concentrating photovoltaic/thermal (CPVT) systems," *Energy Convers. Manag.*, vol. 48, no. 9, pp. 2481–2490, 2007, doi: 10.1016/j. enconman.2007.04.004.

[17] M. I. P. Benjwal, J.S.khan, "Modulation and Simulation of Renewable Energy Source using MPPT Techniques," *Int. J. Adv. Res. Electr. Electron. Instrum. Eng.*, vol. 04, no. 07, pp. 5893–5902, 2015, doi: 10.15662/ ijareeie.2015.0407013.

[18] R. Nasrin, M. Hasanuzzaman, and N. A. Rahim, "Effect of high irradiation and cooling on power, energy and performance of a PVT system," *Renew. Energy*, vol. 116, pp. 552–569, 2018, doi: 10.1016/j.renene.2017.10.004.

Chapter 11

Model Reference Adaptive Control of Solar Photovoltaic Systems: Application to a Water Desalination System

Abderrahmen Ben Chaabene and Khira Ouelhazi

Abstract

The major problem of the industrial sectors is to efficiently supply their energy requirement. Renewable energy sources, in particular solar energy, are intermittently accessible widely around the world. Photovoltaics (PV) technology converts sunlight to electricity. In this work, we present a contribution dealing with a new mathematic development of tracking control technique based on Variable Structure Model Reference Adaptive Following (VSMRAF) control applied to systems coupled with solar sources. This control technique requires the system to follow a reference model (the solar radiation model) by adjusting its dynamic and ensuring the minimal value of error between the plant dynamics and that of the reference solar radiation model. This chapter provides a new theoretical analysis validated by simulation and experimental results to assure optimum operating conditions for solar photovoltaic systems.

Keywords: optimization, solar energy applications, following control, photovoltaic, reference model-forecasting

1. Introduction

Using solar energy in several applications becomes more and more interesting in order to minimize the cost of production for any system coupled with solar photovoltaic energy. In particular, water desalination systems, which we are going to mention in this book chapter, actually require renewable energy sources to minimize the cost of producing huge amounts of needed water.

Solar energy is only available during daylight hours. Furthermore, problems of intermittent natural solar energies are rarely discussed in practical use cases as power fluctuates over time. Thus, we must introduce real-time operating procedures.

In addition, sunrise and sunset cause daily fluctuations, so the energy delivered by panels will not be constant and it can also suddenly vary due to clouds. Thus, there is a risk of not supplying the energy demand of the load. As an example, when the load consists of a water desalination system, the minimum of sunshine variation can provoke the clogging of membranes used for reverse osmosis desalination, which causes the destroying of the whole desalination system.

IntechOpen

The production of needed water in large quantities demands excessive consumption of energy. Thus, we need to reduce the cost of water production to the maximum possible. Many studies have been carried out to link renewable energy with water production [1, 2]. Furthermore, process control is an important part of the coupling of solar energy to industrial systems; in particular, the desalination industry requires to be operated at the optimum conditions. When coupled to solar photovoltaic sources [3], all industrial systems must be equipped with a regulation energy system to guarantee a continuous energy supply.

The intermittence of solar energy provokes an unstable electrical supply of the photovoltaic solar plants, so the industrial system's parameters change their values with time in an unknown range.

The main idea is to develop a control strategy based on following control theory in order to force the industrial system to adapt to variations in solar photovoltaic energy.

Hence, our idea is to combine two types of robust controls: the adaptive control characterized by its real-time adjustment and the sliding mode control, characterized by its robustness [4–9]. Many formulae have been derived to tune the variable Structure Model Reference Adaptive Following control [10].

To guarantee the stability of the photovoltaic system in the opposite of the intermittence of sunshine, we have chosen a new control algorithm designed by VSMRAF judges effective against stochastic disturbances of uncertainties. These uncertainties affecting the dynamic of photovoltaic solar sources come from various sources [11]. Essentially, they are due to the following reasons: solar sources modeling, intermittent sunshine, solar systems position, approximations in dynamic models. Consequently, all these parameters must be taken into account in the development of photovoltaic solar sources control laws in order to optimize their operation and to avoid all the storage disturbances that may take place and affect the performance of the solar-photovoltaic source.

To overcome renewable energy variability, many strategies were proposed such as:

The application of Maximum Power Point tracking (MPPT) has different algorithms. This method requires high-performance control and seems to be heavy in experiment plants adding to high dependency on specialized and accurate sensors (voltage and current sensors) [12].

The theory of Large Law Number (LLN) can be applied in the use of many solar sources to assure the stability of the electric energy offered by the solar source. Obviously, this solution is expensive at the level of the installation of the solar system.

The use of hybrid renewable systems, for example, wind-solar systems is also proposed as a solution to overcome the problem of solar intermittency. Although this technique is practically possible, the geographical diversity of locations represents a real challenge among many other enormous challenges for this solution.

In other literature [13, 14], we proposed several methods of prediction of solar energy and how much it can be available a day. This theory seems to be not practical and extremely difficult since there are several parameters involved in the prediction algorithms.

In this book chapter, we propose a novel technique depending on a dynamic model of the system itself and taking into account the uncertainties introduced by the intermittence of the photovoltaic solar source. This technique reinforces the robustness and the stability of the system with respect to the disturbances caused by the variation of the solar source.

This book chapter is organized as follows: a state of the art has been presented in the introduction to explore the literature and the research developed to solve the problem of solar energy variation and its effect on industrial systems coupled to photovoltaic sources.

The second section is devoted to the material and methods section. We have presented theoretical elements and mathematic development of the VSMRAF control algorithm. This algorithm is applied to an uncertain dynamic model of the system coupled to a photovoltaic solar source. During the coupling of this system to the photovoltaic solar energy, this model will allow us to avoid the disturbances due to the variation of the electric power supply during the intermittence of the solar energy. We have also presented a background on photovoltaic solar sources and the description of our experimental study plant, which consists of a Reverse Osmosis Desalination (ROD) system coupled to a photovoltaic solar source (PV-ROD) system.

The last section deals with the main simulated and experimental results obtained by our experimental plant relating to a real example of coupling a water desalination system to a solar photovoltaic system.

In conclusion section, we have summarized the main results of this book chapter and we have presented the main findings as well as the perspectives of this research work.

2. Material and methods

2.1 Theoretical elements

2.1.1 Multi input multi output linear uncertain systems

The following model represents the state-space model of an uncertain system [15–18].

The pair of matrices (ΔA, ΔB) represents the incertitude affecting the state matrix A and the control matrix B of the linear system.

$$\dot{x} = (A + \Delta A)x + (B + \Delta B)u \tag{1}$$

The uncertainty state matrices of the system correspond to Eq. (1).
where

$$[\Delta A \quad \Delta B] = D_1 \nabla [E_1 \quad E_2] \tag{2}$$

$D_1 \in \Re^{nxd} \; E_1 \in \Re^{exn} \; E_2 \in \Re^{exm}$
The uncertain matrices ΔA and ΔB are bounded in norm.

2.1.2 Novel mathematic development for the following control strategy

In this approach, we will develop the mathematical formulations detailed in [19] concerning the structure of the adaptive sliding mode control of multivariate systems by tracking a reference model, by replacing the multivariate system represented in the state space by Eq. (3).

$$\dot{x} = Ax + Bu \tag{3}$$

If we replace A and B with their novel expressions taking into account the uncertain matrices, we find the Eq. (4).

$$\dot{x} = (A + \Delta A)x + (B + \Delta B)u \tag{4}$$

Our following theorem shows that if we select a reference model, we can neglect the incertitude on ΔB matrix.

2.1.3 Ben Chaabene's theorem

As we have developed in [19], the reference trajectory in the state space Cr is independent of the effect of uncertainties on the system's dynamic. To take into account the uncertainties on the system, we have assimilated its trajectory in the state space to a cylindrical envelope (E_s) with a radius r depending on uncertain matrix ΔA. This matrix depends on the following error and represents the possible deviation of the system evolved from its reference trajectory. As in the VSMRAF control, the reference matrix of control B_r has constant values, so $B_r = B$, and consequently $\Delta B = 0$.

The trajectory can be represented by **Figure 1**.

Consider the state representation of an uncertain system and from the theorem and the Eq. (4), we find that:

$$\dot{x} = (A + \Delta A)x + Bu \tag{5}$$

The state-space reference model is represented by the following equation.

$$\dot{x}_r = A_r x_r + B_r u_r \tag{6}$$

Two matrices Θ^* and Q^* are defined in order to determine the reference model matrices A_r and B_r:

$$A_r = A + \Delta A + B\theta^* \tag{7}$$
$$B_r = BQ^* \tag{8}$$

With:
Θ^*: matrix with dimension (m × n).
Q^*: diagonal matrix (m × m).
The control law stretching the error to zero is defined.

$$u = \psi x + Q^* u_r \tag{9}$$

The matrix Ψ is an against-reaction matrix of dimension (m × n). The switching functions Ψ_{ij} of the matrix Ψ are adjusted by a variable structure approach. The tracking error x_{ei} for the uncertain system is determined as follows:

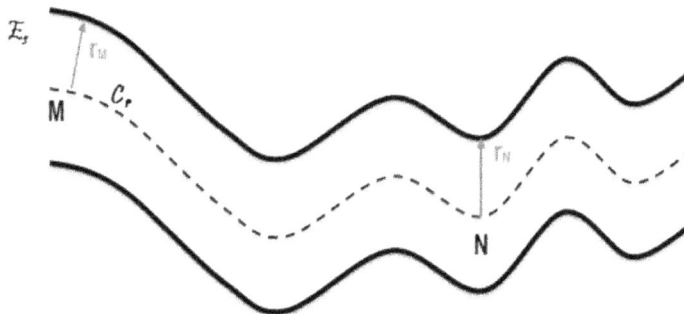

Figure 1.
Effect of the uncertainties on the evolution of the continuation error.

$$x_{ei} = x - x_r \tag{10}$$

If we derive the Eq. (10) with respect to time we find.

$$\dot{x}_{ei} = \dot{x}_i - \dot{x}_{rm} \tag{11}$$

By replacing $\dot{x}$ and $\dot{x}_r$ by their expressions in Eqs. (5) and (6), we obtain:

$$\dot{x}_{ei} = (A + \Delta A)x + Bu - (A_r x_r + B_r u_r) \tag{12}$$

The Eq. (7) gives:

$$A = A_r - \Delta A - B\theta^* \tag{13}$$

Consequently.

$$\dot{x}_{ei} = (A_r - B\theta^*)x + Bu - A_r X_r - B_r u_r \tag{14}$$

By replacing the control law with its expression in Eq. (9) and using the Eq. (14), we obtain:

$$\dot{x}_{ei} = A_r(x - x_r) + B(\psi - \theta^*)x \tag{15}$$

By comparing the Eq. (15) to expressions in literature such as detailed in Eq. (14) for the certain systems (or systems without uncertainties), we find that the derivative of the state error for the uncertain system has the same expression than the following Eq. (16).

$$\frac{de}{dt} = A_r e + B(\psi - \theta^*)X \tag{16}$$

Consequently, the VSMRAF technique allowed us to eliminate the affection of the system dynamic by uncertainties due to sunshine intermittence.

2.2 Application to water desalination system fed by photovoltaic solar source

2.2.1 Design of the experimental plant

The experimental plant that we have used to validate theoretical results is shown in **Figure 2**. It is essentially composed by 3 principal parts:

- A Photovoltaic (PV) system

- High Pressure (HP) pump

- A reverse osmosis desalination (ROD) unit

The brackish or sea water is pumped into a closed vessel and pressurized to the RO unit by a High Pressure (HP) pump, which is fed by the photovoltaic solar source using solar panels. The RO unit is a semi-permeable polyamide membrane, composed of two sides: a brine side and a permeate side. Saltwater is pumped back to the membrane where the salt solution is rejected by the brine side and the desalted pure water passes through the permeate side.

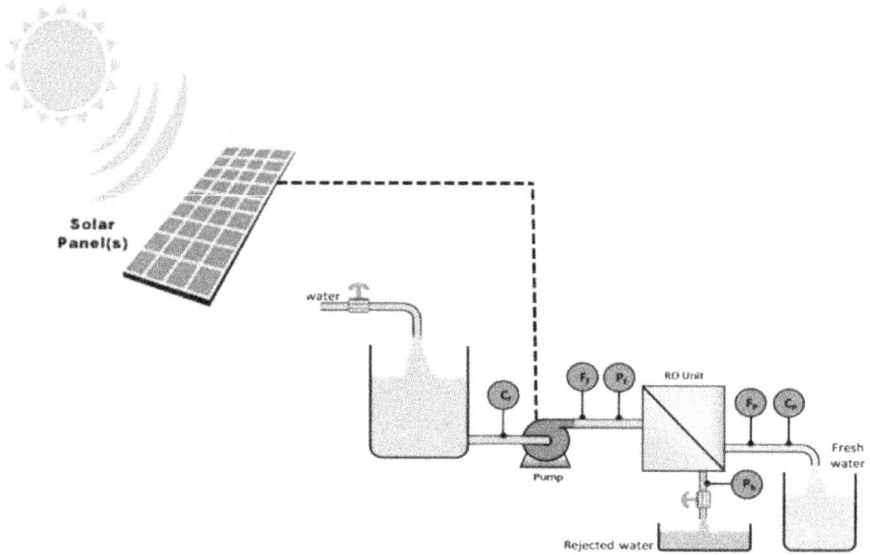

Figure 2.
Design of the photovoltaic reverse osmosis water desalination system.

2.2.2 The photovoltaic system

Solar PV systems are composed of the following components [20]:

- PV arrays consisting of a number of PV modules or panels wired in parallel and/or series association to provide desired voltage and current. Each PV module is composed of many PV cells associated in series to produce high voltages and in parallel to increase current intensity.

  Solar cells are semiconductors made from silicon (single or polycrystalline) devices that convert sunlight into direct current (DC). **Figure 3** shows the equivalent electric circuit of the PV solar cell [21] with all parameters designed in the abbreviations table.

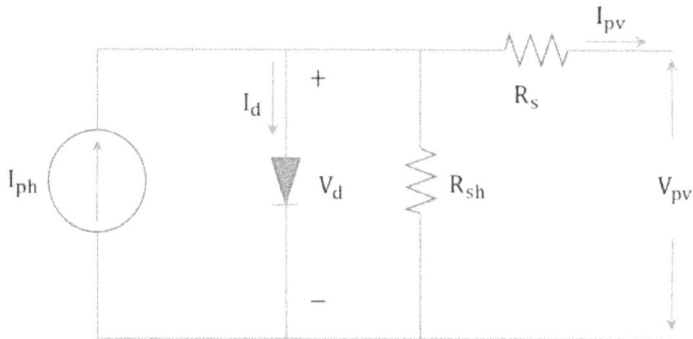

Figure 3.
The equivalent electric circuit of the photovoltaic solar cell.

- Solar inverter, which transforms direct current produced by PV modules to alternative current (AC) needed by AC loads.

- A bias of system (BoS), the generation of AC and DC power.

- Batteries were used in case of off-grid PV systems or when power is needed at night.

Mathematic laws of the PV solar cell model are given by the following expressions detailed in [15].

$$V_D = V_P + R_S I_P \tag{17}$$

$$I_P = I_{ph} - I_D - (V_D/R_{sh}) \tag{18}$$

$$I_P = I_{ph} - I_s \left[\exp\left(q V_D/AKT\right) - 1 \right] - V_D/R_{sh} \tag{19}$$

2.2.3 Description of our photovoltaic system

Figure 4 shows a PV (crystalline silicon) system with 42 kwp capacity, installed at the Research Center of Energy Technologies (CRTEn) in the Borj Cedria techno pole in the south of Tunis.

The installed PV system consists of three PV arrays including 172 modules (64 each one) and 3 inverters with capacity of 17.5 kVA each one. **Table 1** gives the characteristics of one PV module.

2.2.4 Experimental plant

The whole reverse osmosis (RO) desalination system used as an experimental plant is shown as follows. This system contains essentially 3 RO modules, a Moto pump, and an electronic card for data acquisition from various sensors already installed. This system is coupled to a photovoltaic system.

Figure 4.
The photograph of the PV system installed at CRTEn Borj cedria in the south of Tunis.

Characteristics	Units	STC conditions
Maximum power P_{max}	W	250
Voltage at P_{max}	V	28.90
Current at P_{max}	A	8.66
Open circuit voltage	V	37.60
Short circuit current	A	9.29
The total amount of energy produced annually by the PV system is 94.124 MWh/year.		

Table 1.
PV module electrical characteristics.

3. Simulation and experimental results

3.1 Desalination system state-space model

The simulation of the Reverse Osmosis Desalination system dynamic was effected Using Matlab software. The state-space model of the system is given by Eq. (5). The following constant matrices A, B, and C are determined from experimental results.

$$A = \begin{pmatrix} -1 & 1 & 0 & 0 \\ -2.25 & -1.50 & 0 & 0 \\ 0 & 0 & -1 & 1 \\ 0 & 0 & -4.62 & -3.23 \end{pmatrix} \quad B = \begin{pmatrix} 2.50 & 0 \\ 0 & -0.56 \\ 0 & -0.20 \\ -0.81 & 0 \end{pmatrix} \quad C = \begin{pmatrix} 1 & 0 & 0 & 0 \\ 0 & 0 & 1 & 0 \end{pmatrix}$$

$$\Delta A = \begin{pmatrix} 0.09 & 0 & 0 & 0 \\ 0.81 & 0.78 & 0 & 0 \\ 0 & 0 & 0.09 & 0 \\ 0 & 0 & 1.66 & 1.68 \end{pmatrix}$$

As it is shown by the Matlab Simulink model in **Figure 5**, we have replaced all matrices of the state-space model using their real experimental values to test the performances of the following of the model by the system. Values of the uncertain matrix ΔA show that it is bounded in the norm. The real values of the uncertainty matrix show that the deviation of the system from its reference trajectory is limited and offers a margin of robustness to this system during its dynamic evolution.

3.2 VSMRAF control algorithm

Figure 7 shows the chronology of the calculation steps of the VSMRAF control algorithm. After the calculation of the control law, the following error was decreased, then the system was forced to follow its reference model.
Figure 6 shows the PV-RO water desalination experimental set-up used for real experimentations.
Figure 5 shows the PV-RO desalination system diagram using the Matlab Simulink procedure.

3.3 Test of system tracking dynamics

To test the PV-ROD system dynamics, we proceed to impose the reference model and control the real evolution of the system dynamics compared to the

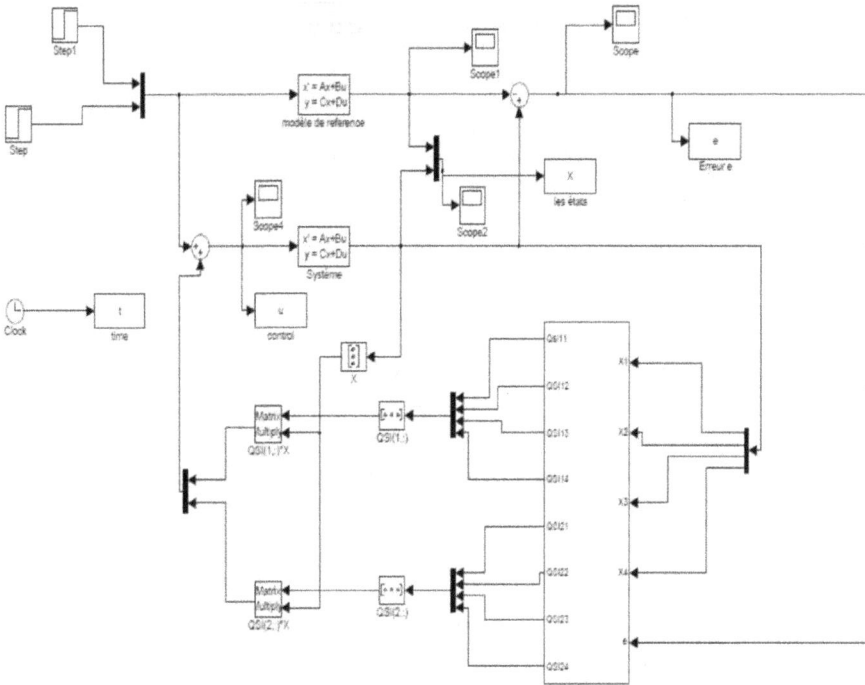

Figure 5.
Matlab Simulink model of the PV-RO desalination system.

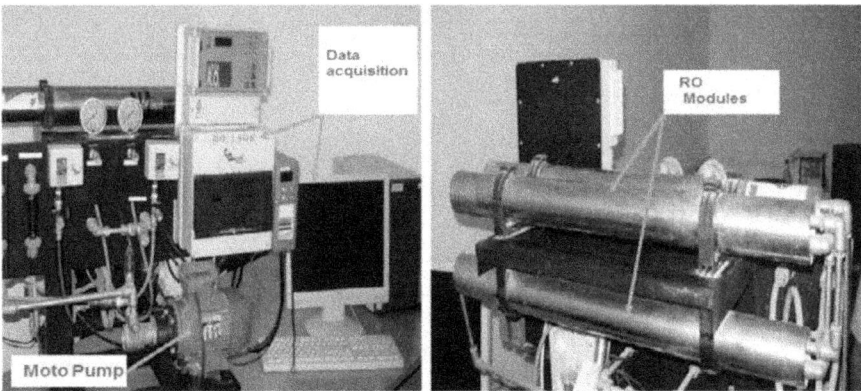

Figure 6.
PV-RO water desalination experimental set-up.

reference model evolution. The following curves show the tracking performance for parametric uncertainties of different values of uncertainties, which correspond to the variations of the photovoltaic solar energy caused by the variation of the sunshine. We note that we have chosen three uncertainty values, which are 10%, 20%, and 40% to follow the tracking performance of the system. The choice of these values of uncertainties stems from the fact that a preliminary study shows that they do not often exceed 25%.

Figures 8 and **9** show respectively, the step responses of the product water flow rate Q and the product salinity Cs for uncertainty of 10%, the reference trajectory is

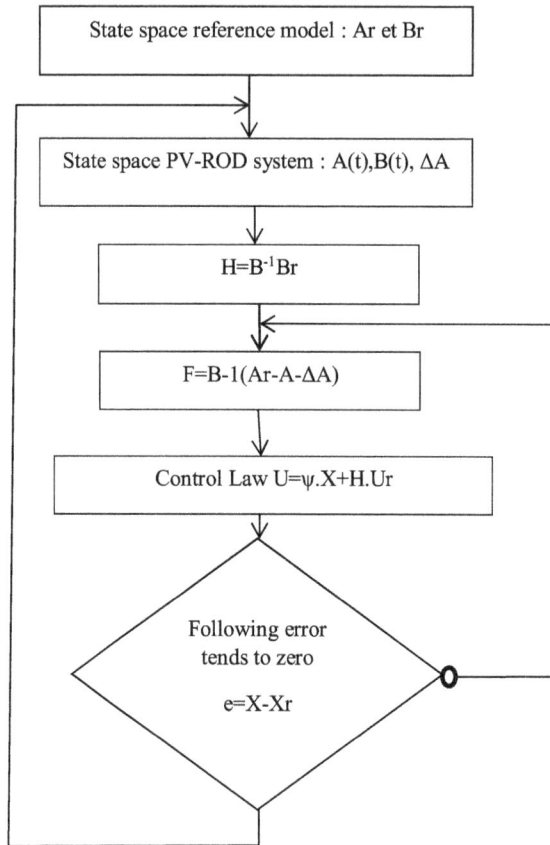

Figure 7.
Algorithm of the VSMRAF control of the PV-ROD system.

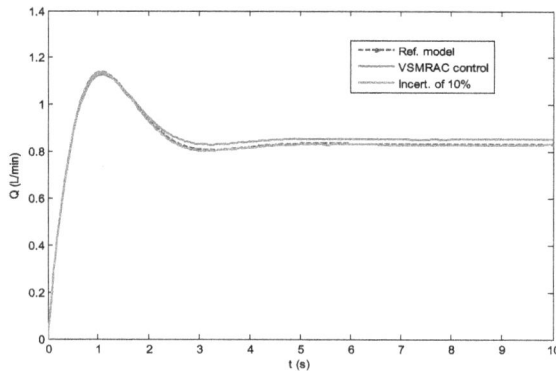

Figure 8.
Flow rate tracking dynamics for uncertainty of 10%.

characterized by a slow dynamic. Curves show perfect model following at a finite time. The tracking error values do not exceed 3% for both of the two outputs Q and Cs.

The evolution of the two parameters Q (flow rate) and Cs (output salinity) is shown in **Figures 10** and **11**. These figures show the perfect following of the model even with an uncertainty of 20%. The tracking error is less than 5%.

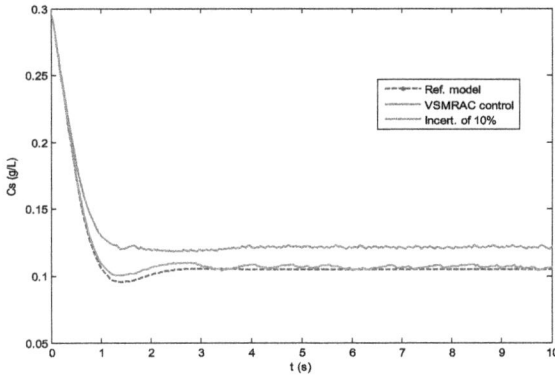

Figure 9.
Salinity tracking dynamics for uncertainty of 10%.

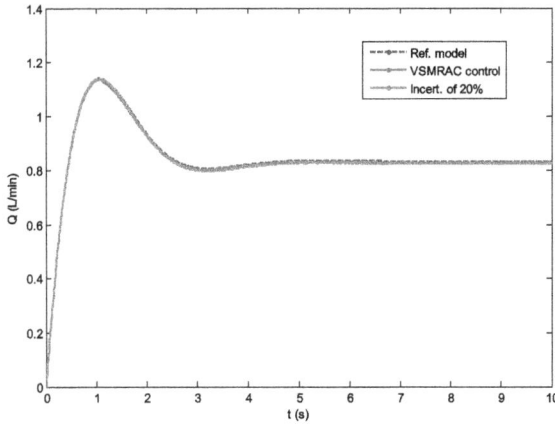

Figure 10.
Flow rate tracking dynamics for uncertainty of 20%.

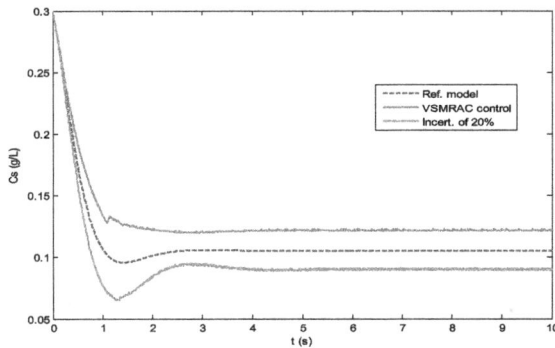

Figure 11.
Salinity tracking dynamics for uncertainty of 20%.

The evolution of the two parameters Q (flow rate) and Cs (output salinity) is shown in **Figures 12** and **13**. These figures show the perfect following of the model even with an uncertainty of 40%. The tracking error is less than 8%.

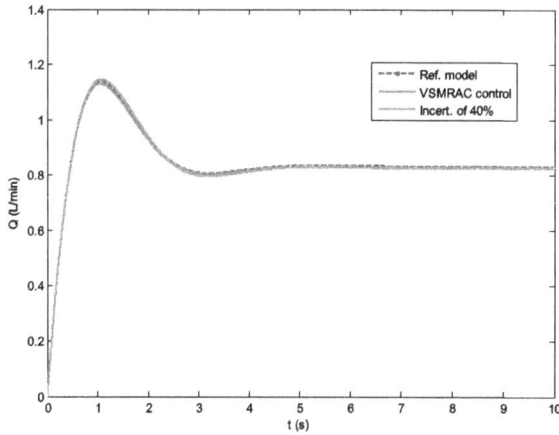

Figure 12.
Flow rate tracking dynamics for uncertainty of 40%.

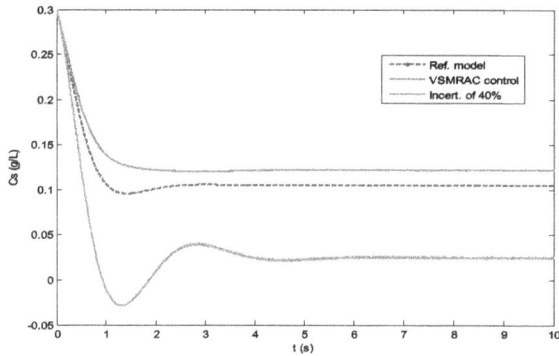

Figure 13.
Salinity tracking dynamics for uncertainty of 40%.

3.3.1 Findings

If we assume that the intermittence of sunshine does not, in any case, exceed 40% of its normal value, the dynamic model of sunshine cannot be more than 40% different from its reference model. Therefore, we can deduce that the dynamic of the system remains insensitive to the variations of the photovoltaic solar source. Thus, we have exceeded the problem of intermittent sunshine, thanks to the implemented VSMRAF control technique.

3.4 Performances of VSMRAF control for optimization of operating systems

From these experimental results, we can essentially deduce that even in the case of high uncertainties, the tracking error remains clearly low, the system remains stable with regard to disturbances due to variations in the sunshine.

Consequently, the control technique that we have developed is suitable for the optimization of operating systems coupled to photovoltaic solar sources. It keeps the stability of the system and eliminates the effect of sunshine variation on its dynamic behavior.

Simulations and experimental results prove our theoretical results and show a perfect following of the reference sunshine model independent of the variation of the photovoltaic solar source. The tracking error is less than 8% remains low even for high intermittence of sunshine not exceeding 40% of its maximum value.

The comparison of results mentioned by the curves for the two output parameters (water flow rate and salinity) of the PV-ROD system shows that the effect of the photovoltaic solar system variation was removed. In addition, experimental results justify the theoretical ones and show that this verified control technique for desalination systems can be generalized for all industrial systems coupled to photovoltaic solar sources having multivariable dynamic models.

4. Conclusion

In this chapter, a VSMRAF Control for systems coupled to photovoltaic solar systems has been proposed. This control technique decreases the sensitivity of the system to variations of solar energy caused by intermittent sunlight. In addition, this technique imposes on the system dynamics to follow the reference model imposed by the daily evolution of solar energy without recourse to the use of batteries.

Furthermore, the experimental and simulation results show that the following error of the system has been kept with a low value (less than 8%) even with a high variation of sunshine. Thus, we can conclude that using this control technique ensures the system's stability and neglects the effect of intermittence of sunshine.

The mathematical development was independent of the nature of the systems, which makes it possible to be applied to all industrial systems, especially those with a limited number of parameters. Obviously, the prospect of this work consists in applying this technique on other multivariable real systems with several parameters and with strongly random variations of the solar energy to better examine the performances of this technique of control.

Abbreviations

I_s	Current of saturation (A)
Q	Charge (C)
V_p	Photovoltaic voltage (V)
V_D	Voltage of diode (V)
R_{sh}	Resistance of cell surface (Ω)
R_s	Resistance of junction face (Ω)
K	Constant of Boltzmann: $1.38064852(79) \times 10$ J–23 J/K.T
T	Temperature (K)
I_p	Photovoltaic current (A)

Author details

Abderrahmen Ben Chaabene* and Khira Ouelhazi
High National School of Engineers of Tunis, Tunis, Tunisia

*Address all correspondence to: abd.chaabene@yahoo.fr

IntechOpen

References

[1] Chaabene A, Elkaroui M, Sellami A. Efficient design of a photovoltaic water pumping and treatment system. American Journal of Engineering and Applied Sciences. 2013;6(2):226-232. DOI: 10.3844/ajeassp.2013.226.232

[2] Chaabene A, Ouelhazi K, Sellami A. Novel design and modeling of a photovoltaic hydro eElectromagnetic reverse osmosis (PV-HEMRO) desalination system. Journal of Desalination and Water Treatment. 2017; 66:36-41. DOI: 10.5004/dwt.2016.0087

[3] Colangelo A, Marano D, Spagna G, Sharma V. Photovoltaic powered reverse osmosis water desalination system. International Journal of Applied Energy. 1999;64:289-305. DOI: 10.1007/978-1-4020-5508-9

[4] Furuta K. VSS type self-tuning control. IEEE Transaction on Industrial Electronics. 1993;40:37-44. DOI: 10.3182/20070829

[5] Mandler J. Modeling for control analysis and design in complex industrial separation and liquefaction processes. Journal of Process Control. 2000;10(2):167-175

[6] Lee H, Utkin V. Chattering suppression methods in sliding mode control systems. IEEE Transaction on System, Man and Cybernetics-Part B: Cybernetic. 2008;38:534-539

[7] Chen X. Adaptive sliding mode control for discrete time multi-input, multi-output systems. Automatica. 2006;42:427435. DOI: 10.1155/2014/673415

[8] Cunha J, Hsu L, Costa R, Lizaralde F. Output feedback model reference sliding mode control of uncertain multivariable systems. IEEE Transactions on Automatic Control. 2003;48(12):2245-2250

[9] Nunes E, Peixoto J, Olivera T, Hsu L. Global exact tracking for uncertain MIMO Linear systems by output feedback sliding mode control. The Journal of The Franklin Institute. 2014; 351(4):2015-2032

[10] Ardeshir K. A variable structure MRAC for a class of MIMO systems. International Journal of Mechanical Systems Engineering. 2007;1(2):4-79

[11] Bajpai P, Dash V. Hybrid renewable energy systems for power generation in stand-alone applications: A review. Journal of Renewable and Sustainable Energy Reviews. 2012;16(5):2926-2939

[12] Ko J, Huh J, Kim J. Overview of maximum power point tracking methods for PV systems in micro grid. Journal of Electronics. 2020;9(5):1-22

[13] Liu Q, Zhang Q. Accuracy improvement of energy prediction for solar-energy-powered embedded systems. IEEE Transactions on VLSI Systems. 2016;24(6):2062-2074

[14] Jebli I, Belouadha FZ, Kabbaj MI, Tilioua A. Prediction of solar energy guided by pearson correlation using Machine Learning. Journal of Energy. 2021;224:109-120

[15] Chaabene A, Sellami A. A novel control of a Reverse Osmosis Desalination system powered by photovoltaic generator. EEE Xplore. 2013;5:1-6. DOI: 10.1109/ICEESA.2013.6578376

[16] Ouelhazi K, Chaabene A, Sellami A, Abdennaceur H. Model Reference Following Control of a UV Water Disinfection System. In: Special Issue - 5th International Conference on Automatic & Signal Processing (ATS-2017). Proceedings of Engineering & Technology (PET). Tunisia: ATS, publisher. 2017;23(2017)

[17] Ouelhazi K, Chaabene A, Sellami A. Design and Control of an Ultraviolet Water Disinfection System Powered by Photovoltaic Source. Journal of Multidisciplinary Engineering Science and Technology. 2017;**4**(9):8203-8210

[18] Ardeshir K. A variable structure MRAC for a class of MIMO systems. International Journal of Mechanical Systems Engineering. 2007;**1**(2):4-79

[19] Chaabene A, Ouelhazi K, Sellami A. Following Control Of MIMO Uncertain Systems, Application to a Water Desalination System Supplied by Photovoltaic Source. Journal of Engineering Technology. Elsevier. 2018; **6**:1-12

[20] Aghaei M, Kumar NM, Eskandarani A, Ahmed H, De Oliviera AKV, Chopra SS. Chapter 5 - Solar PV systems design and monitoring. In: Book Photovoltaic Solar Energy Conversion. 2020. pp. 117-145. DOI: 10.1016/B978-0-12-819610-6.00005-3

[21] Chaabene A, Andolsi R, Sellami A, Mhiri R. MIMO Modeling approach for a small photovoltaic reverse osmosis desalination system. Journal of Applied Fluid Mechanics. 2011;**4**:35-41. DOI: 10.36884/jafm.4.01.11899